PROGRESSIVE COLLAPSE ANALYSIS OF STRUCTURES

PROGRESSIVE COLLAPSE ANALYSIS OF STRUCTURES

Numerical Codes and Applications

DAIGORO ISOBE
University of Tsukuba, Tsukuba, Ibaraki, Japan

ELSEVIER
Butterworth-Heinemann
An imprint of Elsevier

Butterworth-Heinemann is an imprint of Elsevier
The Boulevard, Langford Lane, Kidlington, Oxford OX5 1GB, United Kingdom
50 Hampshire Street, 5th Floor, Cambridge, MA 02139, United States

Library of Congress Cataloging-in-Publication Data
A catalog record for this book is available from the Library of Congress

British Library Cataloguing-in-Publication Data
A catalogue record for this book is available from the British Library

ISBN: 978-0-12-812975-3

For information on all Butterworth-Heinemann publications visit our website at https://www.elsevier.com/books-and-journals

Publisher: Matthew Deans
Acquisition Editor: Ken McCombs
Editorial Project Manager: Peter Jardim
Production Project Manager: Mohanapriyan Rajendran
Designer: Vicky Pearson Esser

Typeset by TNQ Books and Journals

CONTENTS

ABOUT THE AUTHOR

Daigoro Isobe received the degree of Engineering from the University of Tokyo, Japan, in 1994, and is currently a professor at the University of Tsukuba, Japan. He served as a secretary in Computational Mechanics Division of the Japan Society of Mechanical Engineers (JSME) in 2004. He has been a director of the Japan Society for Computational Engineering and Science (JSCES) since 2012, and has contributed to the success of the International Conference on Computational Engineering and Science for Safety and Environmental Problems (COMPSAFE 2014) as a secretary general. He chaired the 3rd International Workshops on Advances in Computational Mechanics (IWACOM-III) in 2015. He currently serves as a director of international exchange activities in JSCES and is also a general council member of the International Association for Computational Mechanics (IACM) since 2013. He received the Kawai Medal from JSCES in 2015.

He is the developer of the ASI-Gauss code and has conducted various structural collapse analyses for over 20 years. For example, the code had been applied to aircraft impact and progressive collapse analysis of the World Trade Center towers in 9/11 incidents, and had succeeded to show the main cause of the high-speed, total collapse phenomena of the towers. These outcomes were scooped up by the media and had influenced the structural design concept of high-rise buildings thereafter. Other investigations include the CTV building collapse in 2011 New Zealand Earthquake and the collapse of the Nuevo Leon buildings in 1985 Mexican Earthquake. The latter was also broadcasted in the NHK TV program. He is now developing an effective planning tool for structural demolition using controlled explosives, and is also working in the field of robotics, applying finite element approaches to dynamic control systems of robots. He has published over 60 refereed papers and six books in these fields. He received the Ichimura Award upon these achievements in structural collapse analysis field, in 2014, in the presence of Princess Akiko of Japan.

PREFACE

As a structural engineer, collapse of structures must be the last thing he or she wants to witness. The total collapse of the World Trade Center (WTC) towers of New York, USA, which happened in 2001, was literally the most shocking incident that occurred in the history of structural engineering. How and why was it possible? Why was the duration of collapse so fast? Many structural engineers in the world must have had the same feeling. Structures originally built to keep people safe and sound, should keep that way. They must NOT collapse and take people's lives in any way. As I recall, the motivation of writing this book must have been brought up from that moment.

Collapse of structures, however, is a very complex phenomenon with strong nonlinearity and various mechanical interactions. It should be thoroughly investigated through experiments and numerical simulations first, to avoid the phenomenon from happening. This book, from that point of view, is aimed to introduce basic theories of a numerical code, which enables one to simulate progressive collapse behaviors of structures. Moreover, some applications of the numerical code, which are arranged in the order of disastrous events, are shown to help the readers understand the validity and utility of the code. The source program of the numerical code is also shown in the appendix, to help students and beginners to understand the basic structure of finite element simulations. This book can also be used as a textbook in "structural analysis," "structural mechanics," "computational mechanics" classes, and so on, which consists of 15 units.

I would like to dedicate this book to all the structural engineers who are struggling every day to build beautiful structures for mankind. I hope this book will help a little for those engineers who are willing to prevent another disastrous event from happening.

Daigoro Isobe
Tsukuba
March, 2017

ACKNOWLEDGEMENTS

The author would like to thank Prof. Yutaka Toi of University of Tokyo, who had inspired and taught the author the essence of the techniques described in this book, when the author was a graduate student. The author would also like to thank the late Prof. Hirohisa Noguchi of Keio University and Prof. Kazuo Kashiyama of Chuo University, who had brought him into core activities of academic societies, where he renoticed the importance of the field he was working in. Many thanks to Prof. Muneo Hori, Prof. Makoto Ohsaki, Prof. Tomoshi Miyamura, Prof. Hiroyuki Tagawa, Prof. Shojiro Motoyui, Prof. Toru Takahashi, Dr. Mika Kaneko, Dr. Tomohiro Sasaki, Dr. Koichi Kajiwara, and Dr. Takuzo Yamashita who have supported me in the E-Simulator project. The author would also like to acknowledge the contributions of numerous students of his lab, of which the number is now counted over 70. Especially, Mr. Masaomi Morishita, Mr. Junichi Tamura, Mr. Satoshi Saito, Mr. Michihiro Tsuda, Mr. Kyaw Myo Lynn, Mr. Kazunori Shimizu, Mr. Kensuke Imanishi, Mr. Masashi Eguchi, Mr. Sion Sasaki, Mr. Tomonobu Omuro, Mr. Naoki Katahira, Mr. Hitoshi Yokota, Ms. Nozomi Kunihara, Mr. Tetsuya Hisanaga, Mr. Takuya Katsu, Ms. Le Thi Thai Thanh, Mr. Yuta Arakaki, Ms. Eri Onda, Mr. Won Sang Han, Mr. Tomoya Ogino, Mr. Ryosuke Negishi, Mr. Zhong Hui He, Mr. Naoki Yokemura, Mr. Yuan Qi Dong, Mr. Kyohei Kuroda, Mr. Takuya Yamamoto, Mr. Jiajie Gu, Mr. Yoshiki Kusaka, Mr. Kenta Takatera, Mr. Hiroaki Ogino, Mr. Masato Katagiri, Mr. Kohei Oi, Mr. Takashi Fujiwara, Mr. Kota Azuma, Mr. Kenta Higashi, Mr. Toshiki Miura, and Mr. Chen Tiansheng have contributed in development of the numerical codes. The contents of this book are mainly composed by their outcomes and the author should emphasize that this book would have never reached to this level without their contribution.

RECOMMENDATION LETTER

At the Rio Olympic Games last summer, the pleasure of the Japanese team who won the silver medal in the 4 × 100 m relay on the ground was a topic all over the world. This is because their achievement exceeds the total sum of the competences of the individual players with a quick and sure baton pass to the following runners.

The first runner in the research and development of the method of progressive collapse analysis of structures described in this textbook was Professor Tadahiko Kawai (University of Tokyo). The rigid bodies-spring model (RBSM) based on his idea widely recognized the validity of discontinuum mechanics models in the highly nonlinear analysis as discussed in this textbook.

The second runner is Professor Yutaka Toi (University of Tokyo) who was advised by Professor Kawai as a graduate student (1974–79). Professor Toi derived the mathematical equivalence conditions for the RBSM and the finite elements, based on which he proposed the adaptively shifted integration (ASI) technique as a combination of the discontinuum mechanics method and the finite element method, establishing the fundamentals of the method.

The third runner is Professor Daigoro Isobe (University of Tsukuba), the author of this textbook. He was taught the basic concepts of the ASI technique and the nonlinear finite element method by Professor Toi as a graduate student (1989–94). Continuing the research energetically, he proposed the more accurate version called the ASI-Gauss technique, and developed the practically effective numerical code with the sophisticated algorithms for member fracture, contact, and separation.

By applying the developed program to the progressive collapse analysis of some actual structures subjected to aircraft impact, fire, blast demolition, seismic loading and debris impact, Professor Isobe has opened up a new way for the simulation of complicated collapse behaviors of large-scale structures. Domestically, its practical importance has already been highly evaluated. The international awareness can also be expected to greatly increase by the publication of this English textbook.

In the present research and development, the inspiration and education for the research successors were successfully conducted through the relation of the advisors and the graduate students, as in the elegant button pass of the Japanese relay team, which has largely contributed to the publication of this excellent textbook. It can be recommended to all engineers, researchers, and students in the field of structural engineering.

Yutaka Toi
Emeritus Professor
University of Tokyo

CHAPTER ONE

Introduction

1.1 AIMS AND SCOPE

Mankind has developed strong and artistic structures over many years. Unfortunately, pleasant occasions with these man-made structures are subjected to change due to natural or man-made disasters. Events such as earthquakes, tsunamis, big fire, terrorist attacks, and poor construction works are a few examples of these disasters. The main aim and scope of this book is to make a contribution to disaster reduction and mitigation. It can be achieved by simulating the collapse behaviors of structures during such disastrous events, by a low-cost, highly accurate, finite-element code which can be used in any low-configuration personal computer.

This book is particularly aimed to explain the basic theories of a numerical code, which enables one to simulate progressive collapse behaviors of structures. Moreover, some applications of the numerical code arranged in the order of disastrous events are also shown to help the readers understand the validity and utility of the code.

Chapter 2 introduces adaptively shifted integration (ASI) technique, which is a basic technique implemented in the numerical code. First, the basic beam theory and incremental finite-element formulation of a linear Timoshenko beam element utilized in this technique is described. Then, the basic theory of the original ASI technique, including a shifting technique of a numerical integration point and its relation to a stress evaluation point, is described. Furthermore, a time-integration scheme for incremental equation of motion based on the updated Lagrangian formulation, along with a formulation to analyze structures under seismic excitation, is also described.

Chapter 3 introduces an ASI-Gauss technique, which is a revised ASI technique, and a main structure utilized in the numerical code for analyzing collapse behaviors of buildings. First, a general idea to increase more accuracy than the original ASI technique is explained in this chapter. Then verification of the code was conducted by comparing the numerical results with the ones obtained by conventional finite-element method (FEM) and the original ASI technique. Furthermore, the results obtained by the ASI-Gauss code was

Progressive Collapse Analysis of Structures
ISBN: 978-0-12-812975-3
http://dx.doi.org/10.1016/B978-0-12-812975-3.00001-3

verified and validated by a detailed analysis performed by a supercomputer and an experiment conducted using a three-dimensional shake table.

Chapter 4 introduces member fracture, contact, and contact-release algorithms which play a key role to simulate the collapse behaviors of buildings. Member fracture was considered by applying the shifting process used in the ASI and ASI-Gauss techniques, and contact and release between members were considered by adopting gap elements and their nodal displacement information to keep the computational cost lower. The member fracture algorithm, an outline of gap elements, flow of elemental contact and release, and simple numerical examples are discussed in this chapter.

Chapter 5 introduces one of the applications of the numerical code; an aircraft impact analysis of the World Trade Center tower, which simulated the events occurred during the 9/11 terrorist attacks. The simulation was conducted to examine the damage and the dynamic unloading phenomena, a so-called "spring-back phenomena," that occurred in the core columns of the towers during impact.

Chapter 6 introduces a fire-induced progressive collapse analysis of high-rise buildings with outrigger truss systems conducted to qualitatively demonstrate the effects of fire and structural parameters on the progressive collapse behaviors of buildings. The effects of fire and structural parameters on the redundant strengths were surveyed by observing the collapse initiation time: the duration from the beginning of the fire until collapse initiation.

Chapter 7 introduces a trial to estimate the risk of progressive collapse of buildings by investigating the relation between a key element index, which indicates the contribution of a structural column to the vertical capacity of the structure, and the scale of progressive collapse. The numerical results on various models with different structural strengths indicated an increase of risk of progressive collapse as the sum of key element index values became larger.

Chapter 8 introduces a blast demolition analysis of buildings, which is another useful application of the numerical code. In this chapter, the code was verified and validated by comparing the predicted result with that of an experimental one. A demolition planning tool utilizing the key element index is also developed and the methods of selecting specific columns to efficiently demolish the whole structure are demonstrated.

Chapter 9 introduces a seismic pounding analysis of adjacent buildings, in which the phenomena are expected to occur mainly during long-period ground motion. An analysis was performed on a simulated model of the Nuevo Leon buildings of Mexico City, which consisted of three similar buildings built consecutively with narrow expansion joints between the buildings, and except one building, collapsed completely in the 1985 Mexican earthquake. It can be seen from the numerical results that the difference of natural periods between the adjacent buildings caused by previous earthquakes may have had triggered the collisions and the collapse.

Chapter 10 introduces a seismic collapse analysis of the CTV building, which collapsed during the Lyttelton aftershock on February 22, 2011, in New Zealand. The result showed its unbalanced strengths in the EW and NS directions because of its biased distribution of anti-seismic walls. Moreover, the collapse behavior was observed with a clear twist-mode vibration around the north wall complex.

Chapter 11 introduces a debris impact analysis of a steel-framed building in tsunami, which was conducted by applying seismic ground motion, fluid forces, and debris collision, continuously in a single simulation. The story drift angle of the building and the drag force applied to the building during the sequence were both investigated and compared between models with and without a wall placed under the water line.

Summaries of the developed numerical code, applications, and the future works are described in Chapter 12.

Appendix A introduces a source program of the ASI-Gauss code to help students and beginners to understand the basic structure of finite-element simulations. Only the essences of the program are shown in the appendix; however, the application files can be downloaded freely from the author's website if required.

Appendix B introduces another version of the ASI technique, which utilizes Bernoulli—Euler beam elements. Although this technique cannot be used in collapse analysis of structures, highly accurate solutions can be obtained with only one element subdivision per member in usual non-collapse analysis. Basic theories with some numerical examples of elastic and elasto-plastic analysis under static and dynamic loadings are shown in this appendix. The code can also be downloaded from the author's website.

Appendix C introduces a different type of collapse analysis, which was not carried out on buildings but on nonstructural components; the ceilings of a gymnasium. According to the numerical result, the collapse of the ceilings progressed owing to the detachment of clips that connected the ceiling joists to the ceiling joist receivers, and eventually led to a large-scale collapse.

Appendix D introduces another application besides collapse analysis of buildings—a motion behavior analysis of furniture during earthquakes. Some furniture were modeled with beam elements and the motion behavior analyses were conducted using the ASI-Gauss code with a contact algorithm based on sophisticated penalty method. The results were compared with the experimental results performed on a shake table.

1.2 DEFINITION AND RECOGNITION OF PROGRESSIVE COLLAPSE

The term *progressive collapse* is defined by Ellingwood et al. (2006) as "Spread of local damage, from an initiating event, from element to element resulting, eventually, in the collapse of an entire structure of a disproportionately large part of it; also known as disproportionate collapse (ASCE 7-05)." The phenomenon was first officially reported when the Ronan Point tower, London, UK, collapsed due to a gas explosion in 1968 (Ministry of Housing and Local Government, 1968). Total collapse of the World Trade Center towers during the 9/11 terrorist attacks in 2001 made the term more popular (ASCE/FEMA, 2002).

1.3 NUMERICAL METHODS TO SIMULATE PROGRESSIVE COLLAPSE BEHAVIORS

The progressive collapse phenomenon is, of course, very difficult to reproduce in experiments. It is also difficult to simulate numerically using computers, since it is packed with many mechanical interactions and strong nonlinear behaviors.

Significant advances in the field of computers are now eliminating the calculation cost restrictions, and various dynamic analysis codes are being developed. Among those codes, there are some codes applicable to dynamic collapse problems which contain strong nonlinearities and discontinuities, such as the distinct element method (Cundall, 1971) and the discontinuous deformation analysis (Shi and Goodman, 1984). These codes have been applied to various demolition analyses and seismic collapse analyses (Tosaka et al., 1988; Yarimer, 1989; Kondo et al., 1990; Meguro and Hakuno, 1991; Itoh et al., 1994; Ma et al., 1995). The applied element method, which can follow the total failure behavior from zero loading to complete collapse in a reasonable CPU time, has also been developed and applied to detailed nonlinear analysis of reinforced concrete structures (Tagel-Din and Meguro, 2000). The main purpose of this book is to implement the new algorithms applicable to such

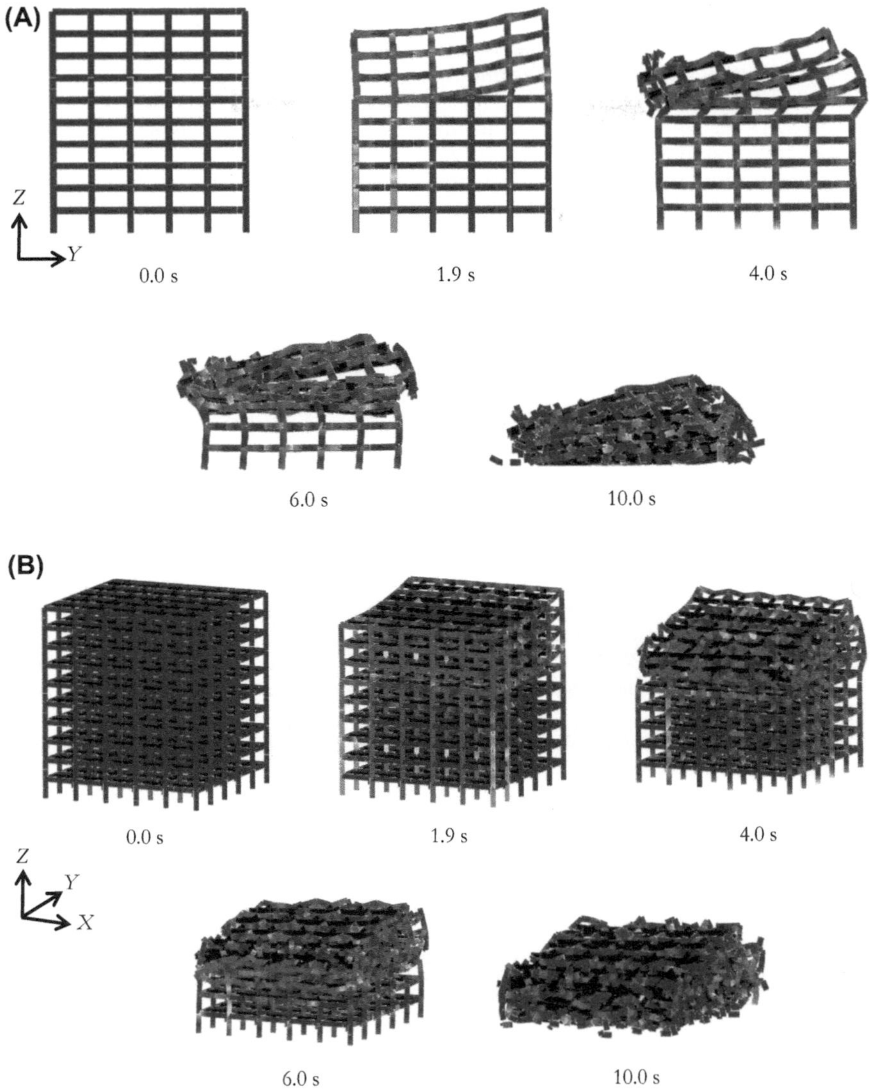

Figure 1.1 A numerical example of progressive collapse analysis. (A) Front view, (B) bird view.

discontinuous problems, which is usually not the main field of the FEM, a more widely used simulation method. By doing so, investigation on progressive collapse behaviors of structures, such as those shown in Fig. 1.1, can be enabled from an analytical point of view.

REFERENCES

ASCE/FEMA, 2002. World Trade Center Building Performance Study: Data Collection, Preliminary Observation and Recommendations. ASCE/FEMA.

Cundall, P.A., 1971. A computer model for simulating progressive, large-scale movement in blocky rock system. In: Proc. International Symposium on Rock Mechanics, pp. 129–136. II-8.

Ellingwood, B.R., et al., 2006. Best Practices for Reducing the Potential for Progressive Collapse in Buildings. National Institute of Standards and Technology (NIST).

Itoh, M., Yoshida, N., Utagawa, N., Kondo, I., 1994. Simulation of blast demolition of reinforced concrete buildings. In: Proc. Third World Congress on Computational Mechanics, pp. 1152–1153.

Kondo, I., Utagawa, N., Ito, M., Yoshida, N., 1990. Numerical method to simulate collapse behavior in blasting demolition of space framed structure. In: Proc. 13th Symposium on Computer Technology of Information, Systems and Applications, pp. 49–54 (in Japanese).

Ma, M.Y., Barbeau, P., Penumadu, D., 1995. Evaluation of active thrust on retaining walls using DDA. Journal of Computing in Civil Engineering 1, 820–827.

Meguro, K., Hakuno, M., 1991. Simulation of collapse process of structures due to earthquake. In: Proc. Symposium on Computational Methods in Structural Engineering and Related Fields, vol. 15, pp. 325–330 (in Japanese).

Ministry of Housing and Local Government, 1968. Report of the Inquiry into the Collapse of Flats at Ronan Point, Canning Town. Her Majesty's Stationery Office, London.

Shi, G.H., Goodman, R.E., 1984. Discontinuous deformation analysis. In: Proc. 25th U.S. Symposium on Rock Mechanics, pp. 269–277.

Tagel-Din, H., Meguro, K., 2000. Analysis of small scale RC building subjected to shaking table tests using applied element method. In: Proc. 12th WCEE, Auckland, New Zealand.

Tosaka, N., Kasai, Y., Honma, T., 1988. Computer simulation for felling patterns of building. In: Demolition Methods and Practice, pp. 395–403.

Yarimer, E., 1989. Demolition by controlled explosion as a dynamical process. In: Structures Under Shock and Impact, pp. 411–416.

CHAPTER TWO

Adaptively Shifted Integration Technique

2.1 INTRODUCTION

A finite-element code to simulate progressive collapse behaviors of structures featured in this book is based on a technique called the adaptively shifted integration (ASI)-Gauss technique (Lynn and Isobe, 2007). It is an improved version of the ASI technique (Toi and Isobe, 1993, 1996, 2000), which is explained in this chapter. There are two versions of the original ASI technique; one utilizing Bernoulli–Euler beam elements (see Appendix B for detail) and another utilizing linear Timoshenko beam elements. Collapse phenomena, however, carry phases such as member yielding and fracture. Fractured section on either side of a member, for example, cannot be considered by a single Bernoulli–Euler beam element with two integration points. A linear Timoshenko beam element with one integration point is more suitable from this point of view. In this chapter, a basic beam theory and an incremental finite-element formulation of the linear Timoshenko beam element are first described. Then, a basic theory of the original ASI technique, including shifting of a numerical integration point and its relation to a stress evaluation point, is described. Furthermore, a time-integration scheme for incremental equation of motion based on the updated Lagrangian formulation (ULF) and a formulation to analyze structures under seismic excitation are also described.

2.2 LINEAR TIMOSHENKO BEAM ELEMENT

A linear Timoshenko beam element is formulated based on the following incremental form of the principle of virtual work:

$$\delta \mathrm{V} - \delta \mathrm{W} = \int_{-^{n}l/2}^{^{n}l/2} \{\delta\Delta_n\varepsilon\}^T\{\Delta_n\sigma\}dz - \{\delta\Delta_n u\}^T\{\Delta_n f\} = 0 \qquad (2.1)$$

where $\delta \mathrm{V}$ and $\delta \mathrm{W}$ are the internal and external works, respectively. $\{\Delta_n\varepsilon\}$, $\{\Delta_n\sigma\}$, $\{\Delta_n u\}$, $\{\Delta_n f\}$, and ^{n}l are the generalized strain increment vector,

Progressive Collapse Analysis of Structures
ISBN: 978-0-12-812975-3
http://dx.doi.org/10.1016/B978-0-12-812975-3.00002-5

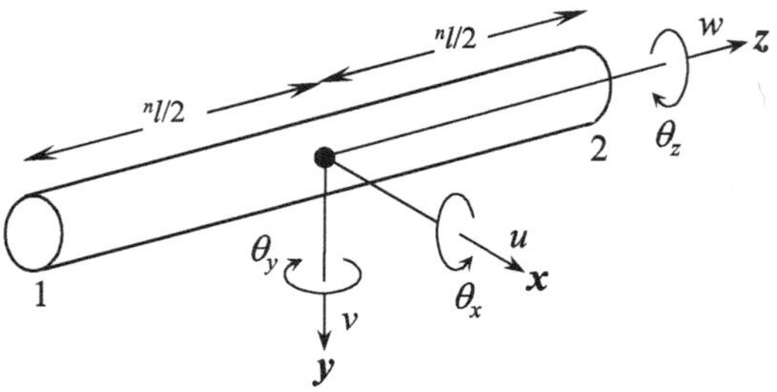

Figure 2.1 Definitions of local coordinates and displacements.

generalized sectional force increment vector, nodal displacement increment vector, external force increment vector, and the length of the element at incremental step n, respectively. The symbols δ and Δ_n denote the variation and increment at step n.

The relation between the generalized strain increment and nodal displacement increment vectors are given by the following equation:

$$\{\Delta_n \varepsilon(r)\} = [B(s)]\{\Delta_n u\} \tag{2.2}$$

where $[B(s)]$ is the generalized strain—nodal displacement matrix; s is the location of the numerical integration point; and r is the location of the stress evaluation point where stresses and strains are actually evaluated. They are nondimensional quantities and take values between -1 and 1. s is defined as $z/(^n l/2)$ in local coordinates shown in Fig. 2.1, and a value of 0 is thoroughly used in conventional FEM. It should be noted that the reduced one-point integration is used for the linear Timoshenko beam element to avoid a "shear locking" mode.

The components of the generalized strain increment vector $\{\Delta_n \varepsilon(r)\}$ at incremental step n are given by the following equations:

$$\Delta_n \kappa_x(r) = -\frac{d\Delta_n \theta_x(z)}{dz} \tag{2.3a}$$

$$\Delta_n \kappa_y(r) = \frac{d\Delta_n \theta_y(z)}{dz} \tag{2.3b}$$

$$\Delta_n \varepsilon_z(r) = \frac{d\Delta_n w(z)}{dz} \tag{2.3c}$$

$$\Delta_n \theta_{z,z}(r) = \frac{d\Delta_n \theta_z(z)}{dz} \tag{2.3d}$$

$$\Delta_n \gamma_{xz}(r) = \frac{d\Delta_n u(z)}{dz} - \Delta_n \theta_y(z) \tag{2.3e}$$

$$\Delta_n \gamma_{yz}(r) = \frac{d\Delta_n v(z)}{dz} + \Delta_n \theta_x(z) \tag{2.3f}$$

where the displacement components (u, v, w, θ_x, θ_y, θ_z) are defined as shown in Fig. 2.1. Each displacement function is defined by the following linear function:

$$u(z) = \frac{1}{2}(1-s)u_1 + \frac{1}{2}(1+s)u_2 \tag{2.4a}$$

$$v(z) = \frac{1}{2}(1-s)v_1 + \frac{1}{2}(1+s)v_2 \tag{2.4b}$$

$$w(z) = \frac{1}{2}(1-s)w_1 + \frac{1}{2}(1+s)w_2 \tag{2.4c}$$

$$\theta_x(z) = \frac{1}{2}(1-s)\theta_{x1} + \frac{1}{2}(1+s)\theta_{x2} \tag{2.4d}$$

$$\theta_y(z) = \frac{1}{2}(1-s)\theta_{y1} + \frac{1}{2}(1+s)\theta_{y2} \tag{2.4e}$$

$$\theta_z(z) = \frac{1}{2}(1-s)\theta_{z1} + \frac{1}{2}(1+s)\theta_{z2} \tag{2.4f}$$

The subscripts 1 and 2 denote the node numbers indicated in Fig. 2.1. By substituting Eq. (2.4) into Eq. (2.3), one can obtain the relations between the generalized strain increments and nodal displacement increments as follows:

$$\Delta_n\kappa_x(r) = -\frac{1}{{}_n l}(\Delta_n\theta_{x2} - \Delta_n\theta_{x1}) \tag{2.5a}$$

$$\Delta_n\kappa_y(r) = \frac{1}{{}_n l}(\Delta_n\theta_{y2} - \Delta_n\theta_{y1}) \tag{2.5b}$$

$$\Delta_n\varepsilon_z(r) = \frac{1}{{}_n l}(\Delta_n w_2 - \Delta_n w_1) \tag{2.5c}$$

$$\Delta_n\theta_{z,z}(r) = \frac{1}{{}_n l}(\Delta_n\theta_{z2} - \Delta_n\theta_{z1}) \tag{2.5d}$$

$$\Delta_n\gamma_{xz}(r) = \frac{1}{{}_n l}\left\{\Delta_n u_2 - \frac{{}_n l}{2}(1+s)\Delta_n\theta_{y2} - \Delta_n u_1 - \frac{{}_n l}{2}(1-s)\Delta_n\theta_{y1}\right\} \tag{2.5e}$$

$$\Delta_n\gamma_{yz}(r) = \frac{1}{{}_n l}\left\{\Delta_n v_2 + \frac{{}_n l}{2}(1+s)\Delta_n\theta_{x2} - \Delta_n v_1 + \frac{{}_n l}{2}(1-s)\Delta_n\theta_{x1}\right\} \tag{2.5f}$$

In summary, the components of the generalized strain increment vector $\{\Delta_n\varepsilon(r)\}$, the nodal displacement increment vector $\{\Delta_n u\}$ and the generalized strain–nodal displacement matrix $[B(s)]$ of Eq. (2.2) are given as follows:

$$\{\Delta_n\varepsilon(r)\} = \{\Delta_n\kappa_x(r) \quad \Delta_n\kappa_y(r) \quad \Delta_n\varepsilon_z(r) \quad \Delta_n\theta_{z,z}(r) \quad \Delta_n\gamma_{xz}(r) \quad \Delta_n\gamma_{yz}(r)\}^T \tag{2.6}$$

$$\{\Delta_n u\} = \{\Delta_n u_1 \quad \Delta_n v_1 \quad \Delta_n w_1 \quad \Delta_n\theta_{x1} \quad \Delta_n\theta_{y1} \quad \Delta_n\theta_{z1} \quad \Delta_n u_2 \quad \Delta_n v_2 \quad \Delta_n w_2 \quad \Delta_n\theta_{x2} \quad \Delta_n\theta_{y2} \quad \Delta_n\theta_{z2}\}^T \tag{2.7}$$

$$[B(s)] = \begin{bmatrix} 0 & 0 & 0 & \frac{1}{_nl} & 0 & 0 & 0 & 0 & 0 & -\frac{1}{_nl} & 0 & 0 \\ 0 & 0 & 0 & 0 & -\frac{1}{_nl} & 0 & 0 & 0 & 0 & 0 & \frac{1}{_nl} & 0 \\ 0 & 0 & -\frac{1}{_nl} & 0 & 0 & 0 & 0 & 0 & \frac{1}{_nl} & 0 & 0 & 0 \\ 0 & 0 & 0 & 0 & 0 & -\frac{1}{_nl} & 0 & 0 & 0 & 0 & 0 & \frac{1}{_nl} \\ -\frac{1}{_nl} & 0 & 0 & 0 & -\left(\frac{1-s}{2}\right) & 0 & \frac{1}{_nl} & 0 & 0 & 0 & -\left(\frac{1+s}{2}\right) & 0 \\ 0 & -\frac{1}{_nl} & 0 & \left(\frac{1-s}{2}\right) & 0 & 0 & 0 & \frac{1}{_nl} & 0 & \left(\frac{1+s}{2}\right) & 0 & 0 \end{bmatrix} \tag{2.8}$$

At those phases when the element behaves elastically, the relation between generalized sectional force increment vector and generalized strain increment vector is given by the following equation:

$$\{\Delta_n\sigma(r)\} = [D_e(r)]\{\Delta_n\varepsilon(r)\} \tag{2.9}$$

where $[D_e(r)]$ is the generalized sectional force–generalized strain matrix for elastic deformation.

The components of the generalized sectional force increment vector $\{\Delta_n\sigma(r)\}$ at incremental step n are given by the following equations:

$$\Delta_n M_x(r) = EI_x\Delta_n\kappa_x(r) \tag{2.10a}$$

$$\Delta_n M_y(r) = EI_y\Delta_n\kappa_y(r) \tag{2.10b}$$

$$\Delta_n N(r) = EA\Delta_n\varepsilon_z(r) \tag{2.10c}$$

$$\Delta_n M_z(r) = GK\Delta_n\theta_{z,z}(r) \tag{2.10d}$$

$$\Delta_n V_x(r) = \alpha_x GA\Delta_n\gamma_{xz}(r) \tag{2.10e}$$

$$\Delta_n V_y(r) = \alpha_y GA\Delta_n\gamma_{yz}(r) \tag{2.10f}$$

where E, G, A, I_x, I_y, and K are the Young's modulus, shear modulus, cross-sectional area, moments of area inertia around x- and y-axes, and Saint-Venant's torsional coefficient, respectively. α_x and α_y are the shear correction factors. In summary, the components of the generalized sectional force increment vector $\{\Delta_n\sigma(r)\}$ and the generalized sectional force–generalized strain matrix $[D_e(r)]$ of Eq. (2.9) are given as follows:

$$\{\Delta_n\sigma(r)\} = \{\Delta_n M_x(r) \quad \Delta_n M_y(r) \quad \Delta_n N(r) \quad \Delta_n M_z(r) \quad \Delta_n V_x(r) \quad \Delta_n V_y(r)\}^T \tag{2.11}$$

$$[D_e(r)] = \begin{bmatrix} EI_x & 0 & 0 & 0 & 0 & 0 \\ 0 & EI_y & 0 & 0 & 0 & 0 \\ 0 & 0 & EA & 0 & 0 & 0 \\ 0 & 0 & 0 & GK & 0 & 0 \\ 0 & 0 & 0 & 0 & \alpha_x GA & 0 \\ 0 & 0 & 0 & 0 & 0 & \alpha_y GA \end{bmatrix} \tag{2.12}$$

It should be noted here that the components of the matrix give constant values for an element. Substitution of Eqs. (2.2) and (2.9) into Eq. (2.1) leads to the following form of elemental stiffness matrix, when inertia terms are neglected:

$$[K_E] = {}^n l[B(s)]^T[D_e(r)][B(s)] \tag{2.13}$$

There are two types of mass matrices for beam elements; lumped mass matrix and distributed mass matrix. The distributed elemental mass matrix for the linear Timoshenko beam element is given as the following equation:

$$[M_E] = \frac{\rho^n l}{6} \begin{bmatrix} 2A & & & & & & A & & & & & \\ & 2A & & & & & & A & & & 0 & \\ & & 2A & & & & & & A & & & \\ & & & 2I_x & & & & & & I_x & & \\ & & & & 2I_y & & & 0 & & & I_y & \\ & & & & & 2I_z & & & & & & I_z \\ A & & & & & & 2A & & & & & \\ & A & & & 0 & & & 2A & & & & \\ & & A & & & & & & 2A & & & \\ & & & I_x & & & & & & 2I_x & & \\ & 0 & & & I_y & & & & & & 2I_y & \\ & & & & & I_z & & & & & & 2I_z \end{bmatrix} \tag{2.14}$$

where ρ and I_z are the density of the member and polar moment of area inertia, respectively. The ASI technique requires less element subdivision, and thus the numerical error tends to increase and becomes apparent if a less detailed, lumped mass matrix is used. Therefore, the distributed mass matrix shown above is adopted to avoid numerical errors produced by lumping each element's mass at each node of an element. Consequently, an implicit scheme should be used for the time integration.

2.3 ADAPTIVELY SHIFTED INTEGRATION TECHNIQUE

In the ASI technique utilizing linear Timoshenko beam element, initial location of the numerical integration point when the element acts elastically is the midpoint of the element; the same location as in the conventional FEM and a suitable location for one-point integration. If a fully plastic section is determined to occur in the element, the numerical integration point is shifted immediately after the determination in order to express a plastic hinge exactly at that section in the element. When the plastic hinge is to be unloaded, the corresponding numerical integration point is shifted back to its initial location.

Fig. 2.2 shows a linear Timoshenko beam element and its physical equivalence to a rigid-body-spring-model (RBSM). As shown in the figure, the relation between the locations of the numerical integration point and stress evaluation point where a plastic hinge is actually formed is expressed as follows (Toi, 1991):

$$s_1 = -r_1 \tag{2.15}$$

As mentioned, the numerical integration point is placed at the midpoint of the element ($s = s_1 = 0$), which gives $r = r_1 = 0$ from Eq. (2.15), when the entire region of an element behaves elastically. Therefore, the

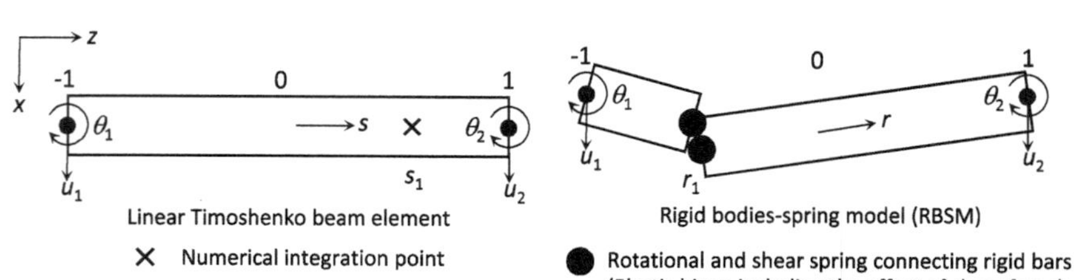

Figure 2.2 Linear Timoshenko beam element and its physical equivalence to RBSM.

elemental stiffness matrix for the elastic element is given by the following equation:

$$[K_E] = {}^{n}l[B(0)]^T[D_e(0)][B(0)] \tag{2.16a}$$

The generalized strain increment vector at $r = r_1 = 0$, $\{\Delta_n\varepsilon(0)\}$, and the generalized sectional force increment vector at $r = r_1 = 0$, $\{\Delta_n\sigma(0)\}$, are calculated as follows:

$$\{\Delta_n\varepsilon(0)\} = [B(0)]\{\Delta_n u\} \tag{2.16b}$$

$$\{\Delta_n\sigma(0)\} = [D_e(0)]\{\Delta_n\varepsilon(0)\} \tag{2.16c}$$

The sectional forces calculated from Eq. (2.16c) are physically ones at the midpoint of the element. According to the beam theory, the relations between the bending moments and shear forces are given as below.

$$V_x = -\frac{dM_y}{dz} \tag{2.17a}$$

$$V_y = -\frac{dM_x}{dz} \tag{2.17b}$$

where M_x and M_y are the bending moments around x- and y-axes, and V_x and V_y are the shear forces along x- and y-axes. Thus, the bending moments along an elastically deformed element can be determined by the following approximations:

$$\Delta_n M_x(s) = \Delta_n M_x(0) - \Delta_n V_y(0)\frac{{}^{n}l s}{2} \tag{2.18a}$$

$$\Delta_n M_y(s) = \Delta_n M_y(0) - \Delta_n V_x(0)\frac{{}^{n}l s}{2} \tag{2.18b}$$

After the formation of a fully plastic section, the numerical integration point is shifted immediately to a new point ($s = s_1 = -r_1$). For example, if a fully plastic section occurs at the right end of an element ($r = r_1 = 1$), the numerical integration point is shifted to the left end ($s = s_1 = -1$) and vice versa. In this case, the elemental stiffness matrix, the generalized strain, and sectional force increment vectors are given as follows:

$$[K_E] = {}^{n}l[B(s_1)]^T[D_p(r_1)][B(s_1)] \tag{2.19a}$$

$$\{\Delta_n\varepsilon(r_1)\} = [B(s_1)]\{\Delta_n u\} \tag{2.19b}$$

$$\{\Delta_n\sigma(r_1)\} = [D_p(r_1)]\{\Delta_n\varepsilon(r_1)\} \tag{2.19c}$$

where $[D_p(r)]$ is the generalized sectional force—generalized strain matrix for plastic deformation. This matrix can be expressed for materials that can be determined by the plastic flow theory, for example, as:

$$[D_p(r_1)] = [D_e(r_1)] - \frac{[D_e(r_1)]\{\partial f/\partial\sigma\}[\partial f/\partial\sigma][D_e(r_1)]}{H' + [\partial f/\partial\sigma][D_e(r_1)]\{\partial f/\partial\sigma\}} \tag{2.20}$$

where H' is the strain hardening coefficient and f is the plastic potential expressed as the following equation:

$$f = f_y(\sigma(r_1)) - 1 = 0 \tag{2.21}$$

It is to be noted that there is no shifting procedure in the mass matrix of the beam elements.

2.4 TIME-INTEGRATION SCHEME FOR INCREMENTAL EQUATION OF MOTION BASED ON THE UPDATED LAGRANGIAN FORMULATION

The incremental equation of motion at time $t + \Delta t$ is given as follows:

$$[M]\{\ddot{u}\}_{t+\Delta t} + [C]\{\dot{u}\}_{t+\Delta t} + [K]\{\Delta u\} = \{F\}_{t+\Delta t} - \{R\}_t \tag{2.22}$$

where $[M]$ is the total mass matrix; $[C]$ is the total damping matrix; $[K]$ is the total stiffness matrix; $\{F\}_{t+\Delta t}$ is the nodal external force vector at time $t + \Delta t$; and $\{R\}_t$ is the nodal internal force vector at time t. If the Newmark's β method is used for time integration, the acceleration and velocity vectors at time $t + \Delta t$ are calculated as follows:

$$\{\ddot{u}\}_{t+\Delta t} = \left(1 - \frac{1}{2\beta}\right)\{\ddot{u}\}_t - \frac{1}{\beta\Delta t}\{\dot{u}\}_t + \frac{1}{\beta(\Delta t)^2}\{\Delta u\} \tag{2.23}$$

$$\{\dot{u}\}_{t+\Delta t} = \left(1 - \frac{\delta}{2\beta}\right)\{\ddot{u}\}_t\Delta t + \left(1 - \frac{\delta}{\beta}\right)\{\dot{u}\}_t + \frac{\delta}{\beta\Delta t}\{\Delta u\} \tag{2.24}$$

where β and δ are the integration parameters. Substituting Eqs. (2.23) and (2.24) into Eq. (2.22) leads to the following equation:

$$\begin{aligned}
&\left(\frac{1}{\beta(\Delta t)^2}[M] + \frac{\delta}{\beta\Delta t}[C] + [K]\right)\{\Delta u\} \\
&\quad = \{F\}_{t+\Delta t} - \{R\}_t + [M]\left\langle\frac{1}{\beta\Delta t}\{\dot{u}\}_t + \left(\frac{1}{2\beta} - 1\right)\{\ddot{u}\}_t\right\rangle \\
&\quad + [C]\left\langle\left(\frac{\delta}{\beta} - 1\right)\{\dot{u}\}_t + \left(\frac{\delta}{2\beta} - 1\right)\{\ddot{u}\}_t\Delta t\right\rangle
\end{aligned} \tag{2.25}$$

where $\{\Delta u\}$ can be treated as the only unknown vector. The calculated results are known to have a numerical damping effect when the range of $\delta \geq 1/2$ and $\beta = (\delta + 1/2)^2/4$ are selected for the parameters (Fung, 2003). The dynamic collapse problems dealt in this book are very strong nonlinear phenomena; thus the values of $\beta = 4/9$ and $\delta = 5/6$ are used for the analyses to stabilize the calculation.

Since extremely large rotations and strains are expected to occur in dynamic collapse analyses, a time-integration scheme based on the ULF is used in the numerical code. The nodal displacement increment vector based on the ULF, at step n, is expressed as follows:

$$\{\Delta_n u\} = [^u T][^0 T]\{\Delta u\} \tag{2.26}$$

where$\{\Delta u\}$ is the initial nodal displacement increment vector; $[^0 T]$ and $[^u T]$ are the transformation matrix from global coordinates to the initial elemental coordinates and the transformation matrix from the initial elemental coordinates to elemental coordinates at step n, respectively. $[^u T]$ is calculated by successive iteration and is expressed as follows:

$$[^u T] = [^n T][^{n-1} T][^{n-2} T]\cdots[^3 T][^2 T][^1 T] \tag{2.27}$$

where $[^n T]$ is the transformation matrix from elemental coordinates at step $n-1$ to elemental coordinates at step n. The matrix $[^n T]$ is calculated as follows:

$$[^n T] = \begin{bmatrix} ^nT^* & 0 & 0 & 0 \\ 0 & ^nT^* & 0 & 0 \\ 0 & 0 & ^nT^* & 0 \\ 0 & 0 & 0 & ^nT^* \end{bmatrix} \tag{2.28}$$

where

$$\begin{aligned} [^n T^*] &= [^n T^\gamma][^n T^\beta][^n T^\alpha] \\ &= \begin{bmatrix} \cos(^n\gamma) & \sin(^n\gamma) & 0 \\ -\sin(^n\gamma) & \cos(^n\gamma) & 0 \\ 0 & 0 & 1 \end{bmatrix} \times \begin{bmatrix} \cos(^n\beta) & 0 & -\sin(^n\beta) \\ 0 & 1 & 0 \\ \sin(^n\beta) & 0 & \cos(^n\beta) \end{bmatrix} \\ &\quad \times \begin{bmatrix} 1 & 0 & 0 \\ 0 & \cos(^n\alpha) & -\sin(^n\alpha) \\ 0 & \sin(^n\alpha) & \cos(^n\alpha) \end{bmatrix} \end{aligned} \tag{2.29}$$

By defining the nodal displacement increments between step $n-1$ and n as $\Delta_{n-1}u_1$, $\Delta_{n-1}v_1$, $\Delta_{n-1}w_1$, $\cdots$, $\Delta_{n-1}\theta_{x2}$, $\Delta_{n-1}\theta_{y2}$, and $\Delta_{n-1}\theta_{z2}$ respectively, $\cos({}^n\alpha)$, $\cos({}^n\beta)$, and ${}^n\gamma$ in Eq. (2.29) can be calculated as follows:

$$\cos({}^n\alpha) = \{{}^{n-1}l + (\Delta_{n-1}w_2 - \Delta_{n-1}w_1)\}/\left[\{{}^{n-1}l + (\Delta_{n-1}w_2 - \Delta_{n-1}w_1)\}^2 + (\Delta_{n-1}v_2 - \Delta_{n-1}v_1)^2\right]^{\frac{1}{2}} \tag{2.30a}$$

$$\cos({}^n\beta) = \{{}^{n-1}l + (\Delta_{n-1}w_2 - \Delta_{n-1}w_1)\}/\left[\{{}^{n-1}l + (\Delta_{n-1}w_2 - \Delta_{n-1}w_1)\}^2 + (\Delta_{n-1}u_2 - \Delta_{n-1}u_1)^2\right]^{\frac{1}{2}} \tag{2.30b}$$

$${}^n\gamma = (\Delta_{n-1}\theta_{z1} + \Delta_{n-1}\theta_{z2})/2 \tag{2.30c}$$

where ${}^{n-1}l$ is the element length at incremental step $n-1$.

The sectional force vector at step $n+1$ can be obtained by transforming the updated Kirchhoff sectional force increment vector $\{\Delta_n\sigma\}$ to the Jaumann differential form vector $\{\Delta_n\sigma^J\}$. The strain vector and sectional force vector are thus expressed as follows:

$$\{{}^{n+1}_{n}\varepsilon\} = \{{}^{n}_{n}\varepsilon\} + \{\Delta_n\varepsilon\} \tag{2.31}$$

$$\{\Delta_n\sigma^J\} = [{}^{n+1}A]\{\Delta_n\sigma\} \tag{2.32}$$

$$\{{}^{n+1}_{n}\sigma\} = \{{}^{n}_{n}\sigma\} + \{\Delta_n\sigma^J\} \tag{2.33}$$

The transformation matrix $[{}^{n+1}A]$ is expressed as

$$[{}^{n+1}A] = \begin{bmatrix} {}^{n+1}T^*_{11} & {}^{n+1}T^*_{12} & 0 & {}^{n+1}T^*_{13} & 0 & 0 \\ {}^{n+1}T^*_{21} & {}^{n+1}T^*_{22} & 0 & {}^{n+1}T^*_{23} & 0 & 0 \\ 0 & 0 & {}^{n+1}T^*_{33} & 0 & {}^{n+1}T^*_{31} & {}^{n+1}T^*_{32} \\ {}^{n+1}T^*_{31} & {}^{n+1}T^*_{32} & 0 & {}^{n+1}T^*_{33} & 0 & 0 \\ 0 & 0 & {}^{n+1}T^*_{13} & 0 & {}^{n+1}T^*_{11} & {}^{n+1}T^*_{12} \\ 0 & 0 & {}^{n+1}T^*_{23} & 0 & {}^{n+1}T^*_{21} & {}^{n+1}T^*_{22} \end{bmatrix} \tag{2.34}$$

where ${}^{n+1}T^*_{ij}$ is the (i, j) term of the $(n+1)$th matrix of Eq. (2.29).

2.5 INCREMENTAL EQUATION OF MOTION FOR STRUCTURES UNDER SEISMIC EXCITATION

In this book, the following incremental equation of motion is used to simulate structures under seismic excitation:

$$[M_1]\{\Delta\ddot{u}\} + [M_2]\{\Delta\ddot{u}_b\} + [C_1]\{\Delta\dot{u}\} + [C_2]\{\Delta\dot{u}_b\} + [K_1]\{\Delta u\} + [K_2]\{\Delta u_b\} = 0 \quad (2.35)$$

The subscript "1" indicates the coupled terms between free nodal points, "2" indicates the coupled terms between free nodal and fixed nodal points, and "b" indicates the components at fixed nodal points. $\{\Delta\ddot{u}\}$, $\{\Delta\dot{u}\}$, and $\{\Delta u\}$ are the nodal acceleration increments, nodal velocity increments, and the nodal displacement increment vectors, respectively.

On the assumption that the displacements at free nodal points are estimated by adding quasi-static displacement increments $\{\Delta u_s\}$ and dynamic displacement increments $\{\Delta u_d\}$, the displacements at free nodal points are given as:

$$\{\Delta u\} = \{\Delta u_s\} + \{\Delta u_d\} \quad (2.36)$$

$\{\Delta u_s\}$ is evaluated by neglecting inertia force, as follows:

$$\{\Delta u_s\} = -[K_1]^{-1}[K_2]\{\Delta u_b\} \quad (2.37)$$

The following equation is obtained by substituting Eqs. (2.36) and (2.37) into Eq. (2.35):

$$[M_1]\{\Delta\ddot{u}_d\} + [C_1]\{\Delta\dot{u}_d\} + [K_1]\{\Delta u_d\} = \left([M_1][K_1]^{-1}[K_2] - [M_2]\right)\{\Delta\ddot{u}_b\} + \left([C_1][K_1]^{-1}[K_2] - [C_2]\right)\{\Delta\ddot{u}_b\} \quad (2.38)$$

In this scheme, equivalent forces are calculated by substituting nodal acceleration increments at fixed points into the right side of Eq. (2.38). Then, the incremental equation of motion is solved by the Newmark's β method.

Generally, it is possible to set the time increment in implicit codes longer than that in explicit codes. Although it is possible to reduce the total computational time by using a longer time increment, the iteration of incremental equations at each step makes the calculation time per step longer, and, in addition, it leads to an excessive use of computer memory. Therefore, the conjugate gradient (CG) method with compressed row storage (CRS) is used as a solver in the numerical code, to reduce the excessive burden on computer memory.

2.6 SUMMARY

Basic theories of the ASI technique utilizing linear Timoshenko beam elements and time-integration schemes used in the dynamic analyses are described in this chapter. The ASI-Gauss technique, an improved version of this technique, is described in the next chapter along with some verification and validation results.

REFERENCES

Fung, T.C., 2003. Numerical dissipation in time-step integration algorithms for structural dynamic analysis. Progress in Structural Engineering and Materials 5, 167–180.

Lynn, K.M., Isobe, D., 2007. Finite element code for impact collapse problems of framed structures. International Journal for Numerical Methods in Engineering 69 (12), 2538–2563.

Toi, Y., 1991. Shifted integration technique in one-dimensional plastic collapse analysis using linear and cubic finite elements. International Journal for Numerical Methods in Engineering 31, 1537–1552.

Toi, Y., Isobe, D., 1993. Adaptively shifted integration technique for finite element collapse analysis of framed structures. International Journal for Numerical Methods in Engineering 36, 2323–2339.

Toi, Y., Isobe, D., 1996. Finite element analysis of quasi-static and dynamic collapse behaviors of framed structures by the adaptively shifted integration technique. Computers and Structures 58 (5), 947–955.

Toi, Y., Isobe, D., 2000. Analysis of structurally discontinuous reinforced concrete building frames using the ASI technique. Computers and Structures 76 (4), 471–481.

CHAPTER THREE

ASI-Gauss Technique

3.1 INTRODUCTION

The general concepts of the ASI-Gauss technique (Lynn and Isobe, 2007), an improved version of the ASI technique utilizing linear Timoshenko beam elements (Toi and Isobe, 1993), are explained in this chapter. In this ASI-Gauss technique, a numerical integration point in an element is shifted immediately after the formation of a plastic hinge and it is shifted back to its initial location when the element is unloaded as in the original ASI technique. The only difference between the ASI and ASI-Gauss techniques lies in their initial locations of the numerical integration points. To compare the numerical accuracies between these methods and the conventional finite-element method (FEM), some verification and validation (V&V) results are shown in this chapter.

3.2 ASI-GAUSS TECHNIQUE

In both the ASI and ASI-Gauss techniques, the numerical integration point is shifted adaptively if a fully plastic section is determined in an element, in order to express a plastic hinge exactly at that section in the element. When the plastic hinge is determined to be unloaded, the corresponding numerical integration point is shifted back to its initial location. By these processes, the plastic behavior of the element is simulated appropriately, and the converged solution can be obtained with a small number of elements per member. Here, the initial location of the numerical integration point is midpoint in the original ASI technique utilizing linear Timoshenko beam element; the location is considered to be optimal for one-point integration when the entire region of the element behaves elastically. However, it causes bending deformations in the elastic range to be inaccurate when the number of elements per member is small, since the displacement functions of the finite element are defined by linear functions. Therefore, a simple means is implemented to improve the accuracy for the elastically deformed member in the ASI-Gauss technique; two consecutive elements forming a member are considered as a subset as shown in Fig. 3.1, and the numerical integration points of an elastically

Progressive Collapse Analysis of Structures
ISBN: 978-0-12-812975-3
http://dx.doi.org/10.1016/B978-0-12-812975-3.00003-7

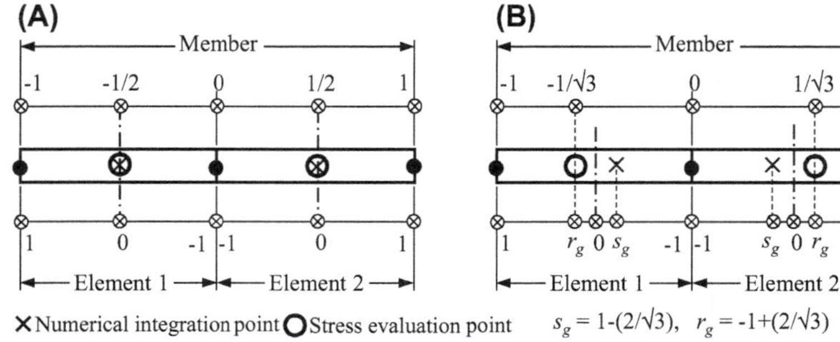

Figure 3.1 Locations of the numerical integration and stress evaluation points in the elastic range. (A) ASI technique and (B) ASI-Gauss technique.

deformed member are placed such that the stress evaluation points coincide with the Gaussian integration points $\left(\pm 1/\sqrt{3}\right)$ of the member, i.e., the stresses and strains are evaluated at the Gaussian integration points of the elastically deformed members. These locations are optimal for two-point integration in the Gaussian quadrature, and the accuracy of the bending deformation defined by a cubic function is mathematically guaranteed. In this way, the ASI-Gauss technique takes advantage of two-point integration while using one-point integration per element in the actual calculations. As in the ASI technique, the relation between the locations of the numerical integration point and the stress evaluation point is expressed as follows (Toi, 1991):

$$s_1 = -r_1 \tag{3.1}$$

As shown in Fig. 3.1, the numerical integration points of the elements in the ASI-Gauss technique are placed at $s = s_1 = s_g$, when the entire region of the member behaves elastically. The initial locations of the numerical integration points and the stress evaluation points in the ASI-Gauss technique are $s_g = 1 - \left(2/\sqrt{3}\right)$ and $r_g = -s_g = -1 + \left(2/\sqrt{3}\right)$, respectively.

Using the concept of ASI technique, the elemental stiffness matrix, generalized strain, and sectional force increment vectors in the elastic range for the ASI-Gauss technique can be given as follows:

$$[K_E] = {}^{n}l[B(s_g)]^T[D_e(r_g)][B(s_g)] \tag{3.2a}$$

$$\{\Delta_n\varepsilon(r_g)\} = [B(s_g)]\{\Delta_n u\} \tag{3.2b}$$

$$\{\Delta_n\sigma(r_g)\} = [D_e(r_g)]\{\Delta_n\varepsilon(r_g)\} \tag{3.2c}$$

where $\{\Delta_n\varepsilon\}$, $\{\Delta_n\sigma\}$, and $\{\Delta_n u\}$ are the generalized strain increment vector, generalized sectional force increment vector, and nodal displacement increment vector at incremental step n, respectively. $[B]$ is the generalized strain–nodal displacement matrix; $[D_e]$ is the generalized sectional force–generalized strain matrix for elastic deformation; and ${}^{n}l$ is the length of the element at incremental step n.

The bending moments along an elastically deformed element can be determined, referring to the same concept as in Eq. (2.18), by the following approximations:

$$\Delta_n M_x(s) = \Delta_n M_x(s_g) - \Delta_n V_y(s_g)\frac{{}^{n}l}{2}(s + s_g) \tag{3.3a}$$

$$\Delta_n M_y(s) = \Delta_n M_y(s_g) - \Delta_n V_x(s_g)\frac{{}^{n}l}{2}(s + s_g) \tag{3.3b}$$

Immediately after the occurrence of a fully plastic section, the numerical integration point is shifted to a new point as in the ASI technique.

The yield condition used for those materials that can be determined by the plastic flow theory, for example, is expressed as follows:

$$f = \left(\frac{M_x}{M_{x0}}\right)^2 + \left(\frac{M_y}{M_{y0}}\right)^2 + \left(\frac{N}{N_0}\right)^2 - 1 \equiv f_y - 1 = 0 \tag{3.4}$$

where f_y is the yield function, and M_x, M_y, and N are the bending moments around the x-axis, y-axis, and axial force, respectively. The terms with the subscript 0 are values that result in a fully plastic section in an element if they act on the cross-section independently.

The elemental stiffness matrix, the generalized strain, and the sectional force increment vectors of the element can be calculated by using Eqs. (2.16) and (2.18) in the ASI technique, and by Eqs. (3.2) and (3.3) in the ASI-Gauss technique when an element behaves elastically. Once a section in an element has yielded based on the yield function, the stiffness matrix and vectors are calculated by using Eq. (2.19) in both techniques. By adaptively shifting the numerical integration point of an element referring to these equations, a precise location of a plastic hinge is found and highly accurate elasto-plastic solutions can be obtained with a very small number of elements.

3.3 VERIFICATION AND VALIDATION OF THE ASI-GAUSS CODE

In this section, some numerical results of elastic and elasto-plastic analyses using a simple space frame are discussed to verify the numerical code utilizing the ASI-Gauss technique. The space frame was subjected to static and dynamic horizontal loads, and to a seismic excitation. In these analyses, the following three methods were applied to verify the accuracies of each method: (1) conventional FEM in which the numerical integration point of each element is fixed at its midpoint, (2) FEM using the ASI technique, and (3) FEM using the ASI-Gauss technique. Furthermore, the numerical code utilizing the ASI-Gauss technique was verified and validated by the numerical results performed by an E-Simulator and a full-scale experimental result obtained at Hyogo Earthquake Engineering Research Center (E-Defense), Japan.

3.3.1 Elastic and Elasto-Plastic Behaviors of a Simple Space Frame Under Static Loading

First, elastic and elasto-plastic behaviors of a simple space frame were analyzed under static loading. Fig. 3.2 shows the analyzed space frame with its dimensions, physical properties, and material constants [this model appears in the input data of ASIFEM (see Appendix A)]. As shown in the figure, the frame was fixed at its lower ends and a static horizontal load was applied to its upper left corner. Members were subdivided into elements according to the numbers specified in the legends of the figures. As large deformations were not expected, a time integration scheme based on the total Lagrangian formulation was used in these analyses.

Figs. 3.3 and 3.4 show the results of the elastic and elasto-plastic analyses under static loading. In conventional FEM, convergence is extremely slow for both the elastic and elasto-plastic analyses. In the elastic analysis, the conventional FEM requires at least eight elements per member to obtain the converged solution. Similarly, over 16 elements per member are required for the elasto-plastic analysis. These phenomena come from the

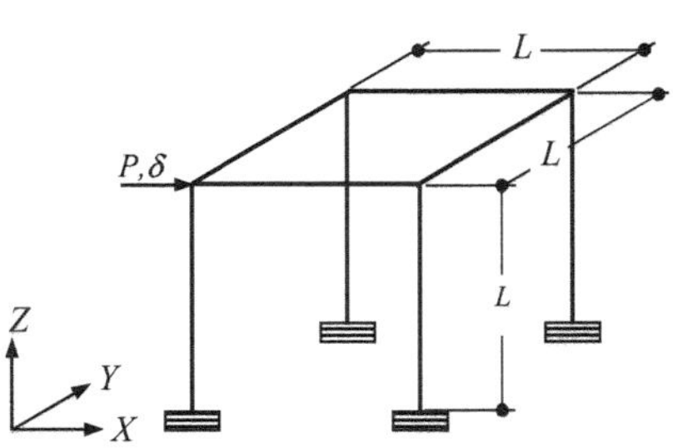

$L = 2.00\ m,\ A = 2.50\times10^{-3}\ m^2$

$I_x = 2.08\times10^{-6}\ m^4$

$I_y = 1.30\times10^{-7}\ m^4$

$E = 206\ GPa, \upsilon = 0.3$

$\sigma_y = 245\ MPa$

$M_{x0} = 1.53\times10^4\ Nm$

$M_{y0} = 3.83\times10^3\ Nm$

$N_0 = 6.13\times10^5\ N$

$\rho = 7{,}900\ kg/m^3$

Figure 3.2 A simple space frame.

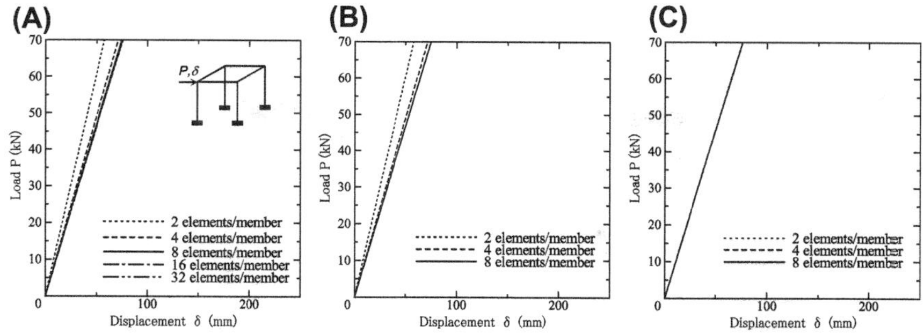

Figure 3.3 Load–displacement relations in elastic analysis. (A) Conventional FEM, (B) ASI technique, and (C) ASI-Gauss technique.

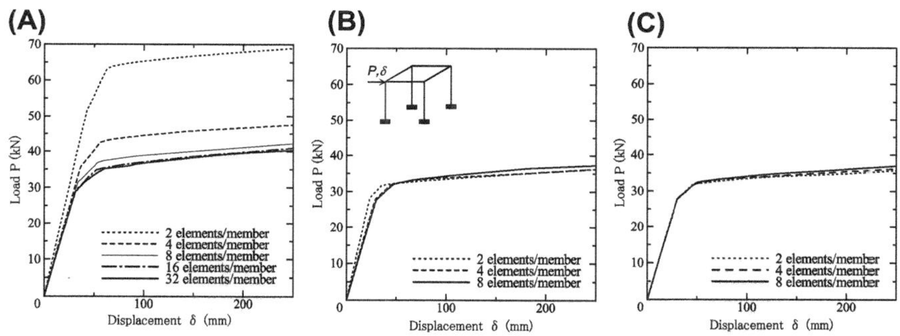

Figure 3.4 Load–displacement relations in elasto-plastic analysis. (A) Conventional FEM, (B) ASI technique, and (C) ASI-Gauss technique.

facts that the displacement functions are defined by linear functions and the numerical integration points are always fixed at the midpoints in the elements. The ASI technique also shows slow convergence for the elastic analysis, in which no plastic section is formed within the elements, and at least eight elements are required for a member to obtain the converged solution—the same situation as in the conventional FEM. Although convergence is slow in the elastic range, the collapse load obtained for the less-element model shows good agreement with the converged solution in the elasto-plastic analysis using the ASI technique; this is enabled by adaptively shifting the numerical integration points to form plastic hinges at appropriate locations. On the other hand, the ASI-Gauss technique shows high convergence in both the elastic and elasto-plastic analyses, and only two-element subdivision per member is necessary to obtain the converged solution; this is enabled by arranging the initial locations of the stress evaluation points in elastic range at optimal locations for two-point integration in the Gaussian quadrature.

3.3.2 Elastic and Elasto-Plastic Responses of a Simple Space Frame Under Dynamic Loading

Elastic and elasto-plastic response analyses were performed later for the same model as shown in Fig. 3.2. The time increment for the dynamic analyses was set to 0.5 ms. A load of 8 kN was applied in the elastic response analysis and 50 kN in the elasto-plastic response analysis. Loading was done for a total of 100 steps (0.05 s) in both analyses.

Figs. 3.5 and 3.6 show the results of the elastic and elasto-plastic response analyses under dynamic loading, respectively. As same as the results provided in the previous section, the conventional FEM shows very slow convergence and over 16-element subdivision is required to obtain the converged solution both in elastic and elasto-plastic response analyses. The results obtained by the ASI technique shows same tendency to the one obtained by the conventional FEM in the elastic response analysis, because there are no differences in the locations of the numerical integration points in the elastic range. On the other hand, the less-element modeling using the ASI-Gauss technique shows a different tendency, as numerical integration points are placed at more suitable locations in the elastic range as described in Section 3.2. In elasto-plastic analysis, the ASI technique gives comparatively better results than the conventional FEM. However, poor convergence can be observed and the vibration mode is much different from that of the converged solution when it comes to the less-element

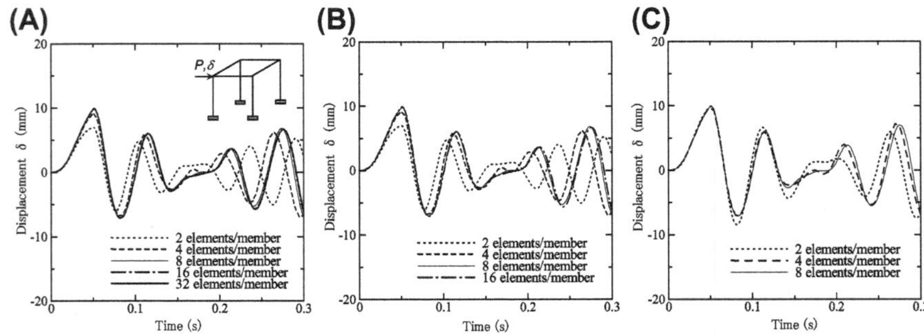

Figure 3.5 Elastic responses of a simple space frame. (A) Conventional FEM, (B) ASI technique, and (C) ASI-Gauss technique.

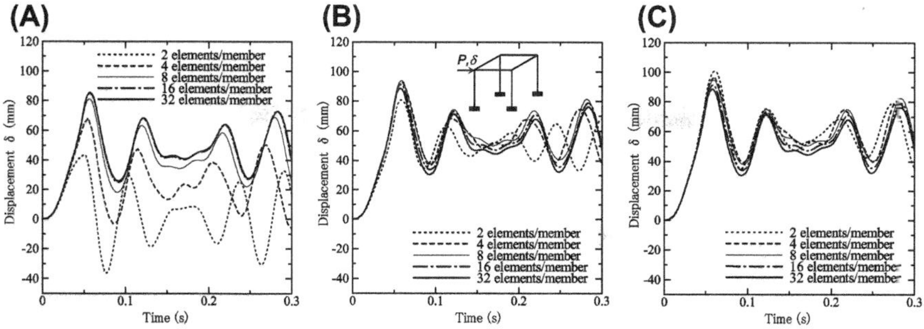

Figure 3.6 Elasto-plastic responses of a simple space frame. (A) Conventional FEM, (B) ASI technique, and (C) ASI-Gauss technique.

modeling. On the other hand, although the results obtained by the ASI-Gauss technique tends to become slightly deformable compared to the exact solution, the convergence is very fast and the solutions are nearly converged even when the number of elements per member is two to four.

From these results, it can be confirmed that the ASI-Gauss technique is an efficient method to analyze static and dynamic behaviors of frame structures both in elastic and elasto-plastic ranges, even by modeling with minimum subdivision per member.

3.3.3 Elastic and Elasto-Plastic Responses of a Simple Space Frame Under Seismic Excitation

Moreover, a 100% EW component of JMA Kobe seismic wave was subjected at the fixed points of the model shown in Fig. 3.2, and elastic and elasto-plastic responses of the frame were observed. In these analyses, the moments of area inertia of the beams were multiplied for over 10^5 times of the original values to reduce unnecessary outer-plane deformations of the upper story, and a floor load of 6000 kgf was subjected. An implicit scheme using Newmark's β method ($\beta = 4/9$ and $\delta = 5/6$ was used to consider numerical damping) and updated Lagrangian formulation were adopted in the analyses.

Fig. 3.7 shows the elastic responses obtained at the upper left corner of the model. The lack of accuracy due to a low-degree displacement function can be noticeably seen in the results obtained by the conventional FEM (Fig. 3.7A) and the ASI technique (Fig. 3.7B). The ASI-Gauss technique, with its numerical integration points at appropriate locations in the elastic range, provides almost exact solution with small element subdivision, as in Fig. 3.7C.

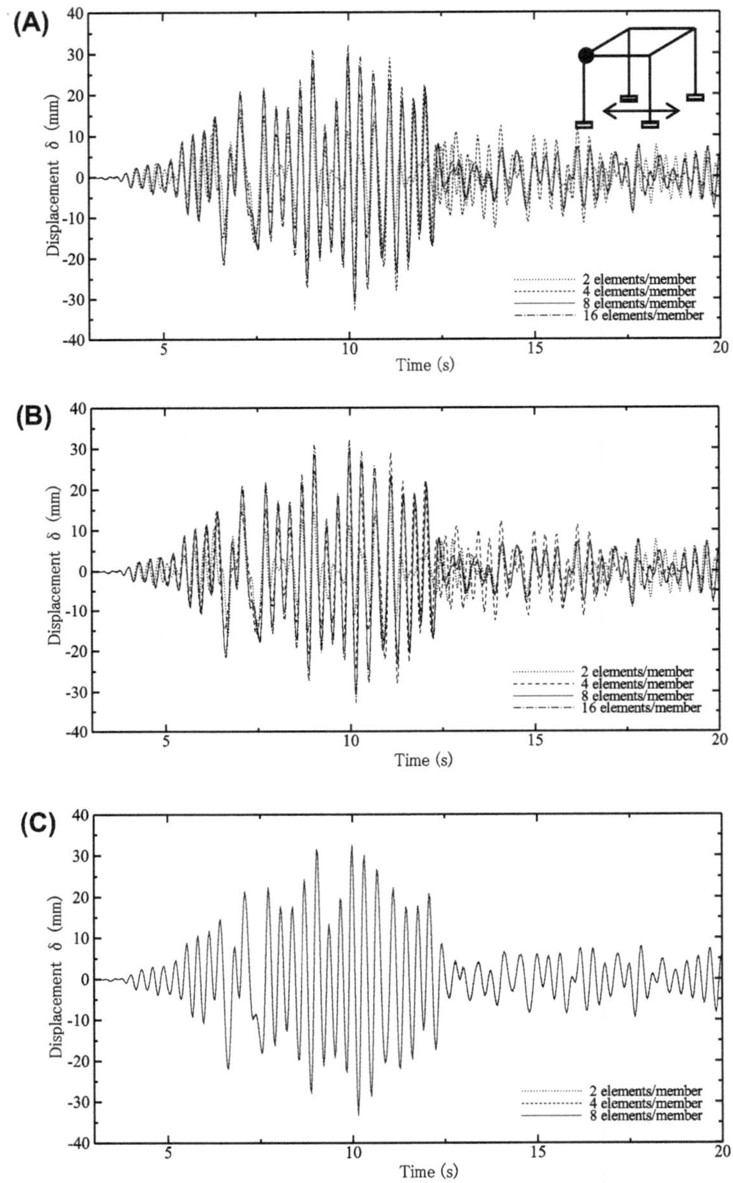

Figure 3.7 Elastic responses of a simple space frame under seismic excitation. (A) Conventional FEM, (B) ASI technique and (C) ASI-Gauss technique.

Fig. 3.8 shows the elasto-plastic responses obtained at the upper left corner of the model. Since plastic hinges are expressed only at the midpoints of the elements, the conventional FEM shows a very slow convergence, also in this case, and it finally converges when 16-element modeling or over are used. The ASI technique gives comparatively better results than the conventional FEM. However, the two- or four-element modeling does not give converged solution, since the elements lack accuracy in the elastic range. On the other hand, the ASI-Gauss technique shows a very fast convergence, and a converged solution is obtained with the minimum subdivision, again, because the stress evaluation points in the elements are adaptively controlled by shifting the numerical integration points in both the elastic and plastic ranges. The three-dimensional excitation gave the same tended results.

3.3.4 Verification and Validation of the ASI-Gauss Code Using Detailed Analysis Performed by E-Simulator and a Full-Scale Experiment Conducted by E-Defense

In this section, verification and validation (V&V) results of the ASI-Gauss code are shown (Isobe et al., 2013). The performance of a seismic response analysis code using the ASI-Gauss technique was verified by comparing the numerical results to those obtained by the E-Simulator, which is a parallel finite-element structural analysis code. Simultaneously, the ASI-Gauss code was validated by comparing the experimental results obtained by a full-scale shake-table test conducted at the E-Defense.

The main core of the E-Simulator is the commercial software package ADVENTURECluster (Akiba et al., 2006), which has been extended from the open source version, ADVENTURE (Yoshimura et al., 2002). The ADVENTURECluster code enables large-scale analyses to be performed with a very fine mesh of solid elements. It has been fully verified by solving various types of benchmark problems in various engineering fields (Yamashita et al., 2012). The experimental result used for validation is that of the full-scale total collapse shake-table test performed on a four-story steel frame (Website of Hyogo Earthquake Engineering Research Center (E-Defense)). Fig. 3.9 shows the specimen of the steel frame, its floor plan and elevations (Website of ASEBI at E-Defense). The member list and the dead load distribution of the building are shown in Tables 3.1 and 3.2, respectively.

Two types of finite-element models were constructed: one for E-Simulator and one for ASI-Gauss code. Fig. 3.10 shows the detailed model of the four-story steel frame constructed for the E-Simulator, where

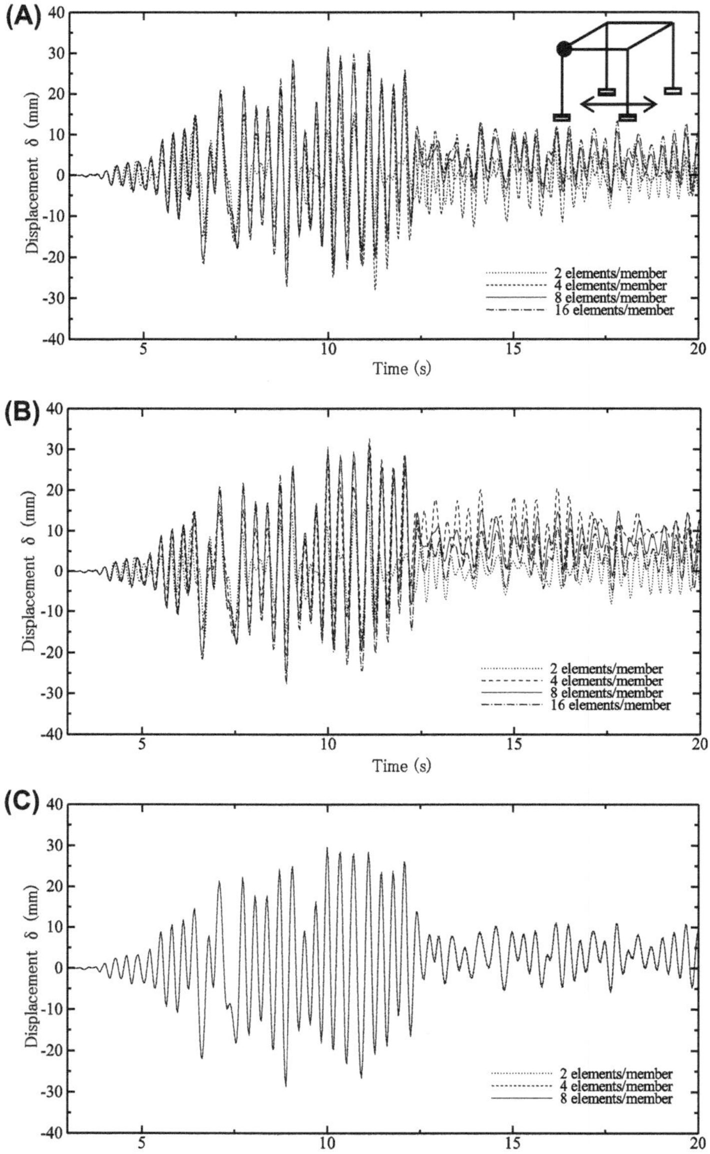

Figure 3.8 Elasto-plastic responses of a simple space frame under seismic excitation. (A) Conventional FEM, (B) ASI technique, and (C) ASI-Gauss technique.

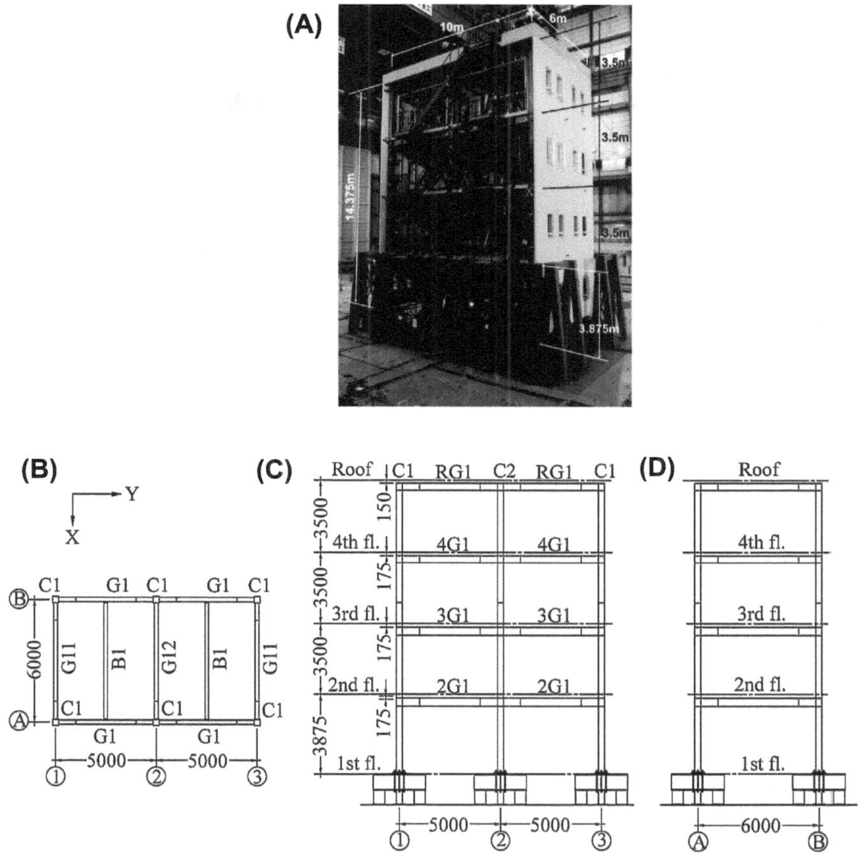

Figure 3.9 Floor plan and elevations of the four-story steel frame (Website of ASEBI at E-Defense). (A) Specimen of the four-story steel frame, (B) plan (first floor), (C) *Y*-elevation, and (D) *X*-elevation.

Table 3.1 Member List of the Four-Story Steel Frame

Member Type	Position and Floor	Section	Material
Column	All	Box-300 × 300 × 9	BCR295
Beam	2G1, 2G11, 3G11, 3G12	H-400 × 200 × 8 × 13	SN400B
	3G1	H-396 × 199 × 7 × 11	SN400B
	2G12	H-390 × 200 × 10 × 16	SN400B
	4G1, 4G11	H-350 × 175 × 7 × 11	SN400B
	RG1, RG11, RG12	H-346 × 174 × 6 × 9	SN400B
	4G12	H-340 × 175 × 9 × 14	SN400B
	B1	H-346 × 174 × 6 × 9	SS400

Table 3.2 Dead Load Distribution of the Four-Story Steel Frame (kN)

	Floor	Steel	Exterior	Partition Wall	Ceiling	Parapet	Protection	Step Floor	Total
Roof floor	459	20			12	71		2	564
Fourth layer		19	79	35					133
Fourth floor	270	24			3		47	4	348
Third layer		18	73	30					121
Third floor	260	32			3		47	4	346
Second layer		18	73	30			8		129
Second floor	260	41					47	4	352
First layer		27	76				12		115
Total	1249	199	301	95	18	71	161	14	2108

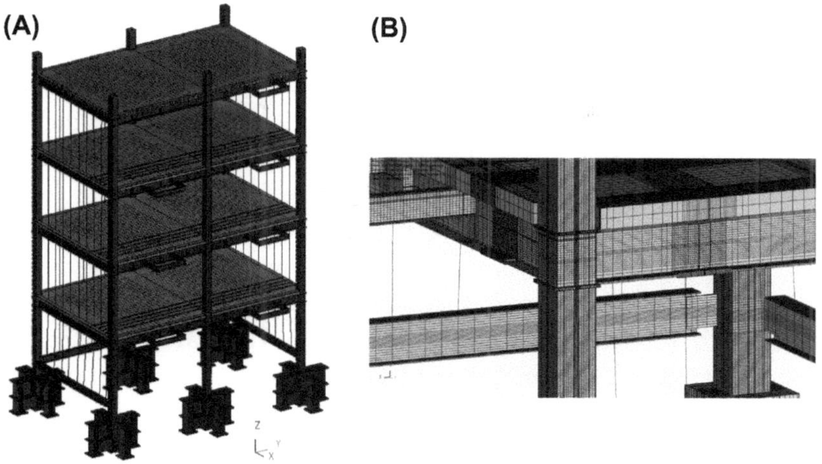

Figure 3.10 Detailed model of the four-story steel frame for the E-Simulator (22.57 million DOFs) (Miyamura et al., 2011). (A) Whole frame and (B) close-up view.

the mesh has 5,181,880 elements, 7,523,255 nodes, and 22.57 million degrees of freedom (DOFs) (Miyamura et al., 2011). The plates, such as the flanges and webs of beams, were divided into at least two layers of solid elements. Each floor slab was also divided into solid elements with two layers. The studs connecting the flanges and the slab were omitted in this model, and the lower surface along the boundary of the slab was directly connected to the upper layer of the flange. The steel bars in the slab were omitted. The size of each element in the longitudinal direction of a beam or a column was approximately 13 mm near the connections, where severe plastic deformation was expected, while a coarser mesh was used for elements located far from the connections. Piecewise linear isotropic hardening was used in the constitutive model of the steel material, and its parameters were determined from the uniaxial test results. A bilinear relationship was used in the constitutive model of the concrete of the floor slab. The self-weight of the steel was calculated based on a mass density of 7.86×10^3 kg/m^3. In contrast, the slab's mass density of 2.3×10^3 kg/m^3 was appropriately increased to include the weights of nonstructural components, anti-collapse frames, and stair landings installed in the experimental model. In the numerical model shown in Fig. 3.10A, the column bases were modeled using solid elements. However, the results shown in this section were obtained using the numerical model in which the column bases were modeled by rotational springs around the *X*- and *Y*-axes whose rotational stiffness was assigned based on the recommendations of the Architectural Institute of Japan (Architectural Institute of Japan, 1990), and the stiffness was evaluated from the anchor bolts in tension. The rotational stiffness around the *Z*-axis of the column base was 10 times as large as the stiffness around the *X*- and *Y*-axes. The Rayleigh damping with coefficients of $\alpha = 0.2284$ and $\beta = 0.001297$ was applied. These coefficients correspond to a damping factor of 0.02 for the first and fourth modes obtained by the eigenvalue analysis, which were the two lowest modes in the *X*-direction.

Fig. 3.11 shows the beam element model for the ASI-Gauss code constructed with 1388 elements, 995 nodes, and 5970 DOFs. Each column and beam in the structure was modeled with two linear Timoshenko beam elements per member. Instead of considering the composite beam effect, the floor slabs in the structure were modeled as a braced truss with stiffnesses and weights approximated to actual values. Offsets of the slabs' positions from the beams were considered by imposing appropriate vertical beams. The structural strengths of the stairs and the exterior walls were ignored, and only their masses were taken into account as a dead load. The bases were completely fixed to the ground, and the stiffness of the base columns was increased appropriately to match the experimental model. In the original calculation process of the ASI-Gauss technique, the bending moments are estimated at the exact location of the nodes. However, the bending moments were estimated at the face of the connections by altering

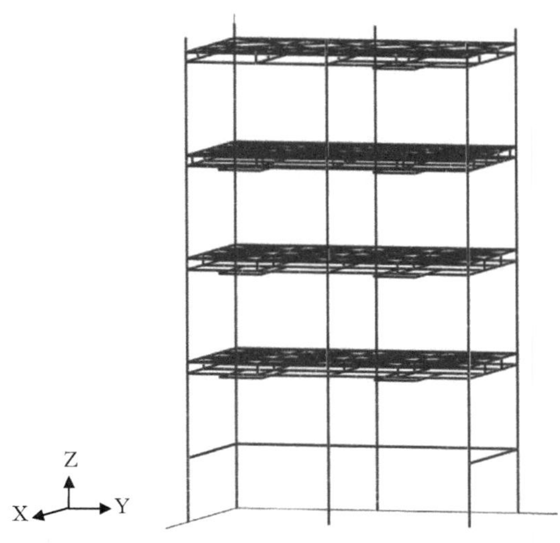

Figure 3.11 Beam element model of the four-story steel frame for the ASI-Gauss code (5970 DOFs).

Eq. (3.3), in this analysis, to reduce the modeling error. Bilinear isotropic hardening was used in the constitutive model of the steel material, and its parameters were determined from uniaxial test results. A mass damping with a coefficient of $\alpha = 0.2284$ was applied.

A 20-second segment of the JR Takatori wave was used as an input ground motion for the experiment performed in the E-Defense. The acceleration measured on the shaking table during the full-scale test (Website of ASEBI at E-Defense) was used for the input ground motion in both analyses. The computational time using the E-Simulator took approximately 1 month using an SGI Altix 256 core computer, and the ASI-Gauss code took approximately 150 min using an Intel Core i7 2.93 GHz personal computer. The time histories of the interstory drift angles and the shear forces of the first story are plotted in Figs. 3.12 and 3.13, respectively.

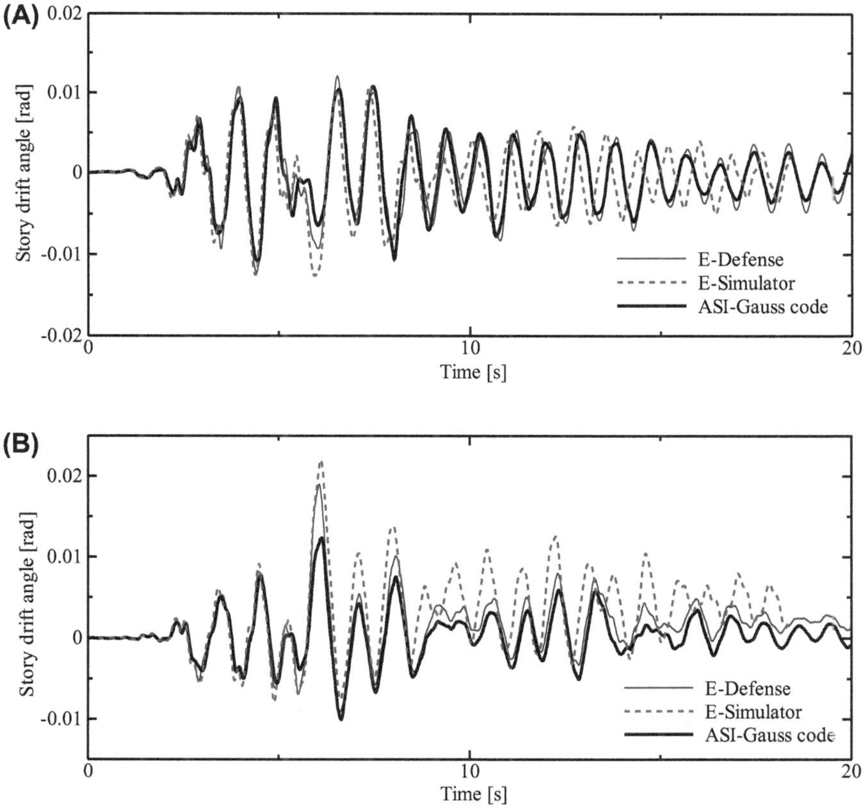

Figure 3.12 Comparison of the time histories of the interstory drift angle on the first floor. (A) *X*-direction and (B) *Y*-direction.

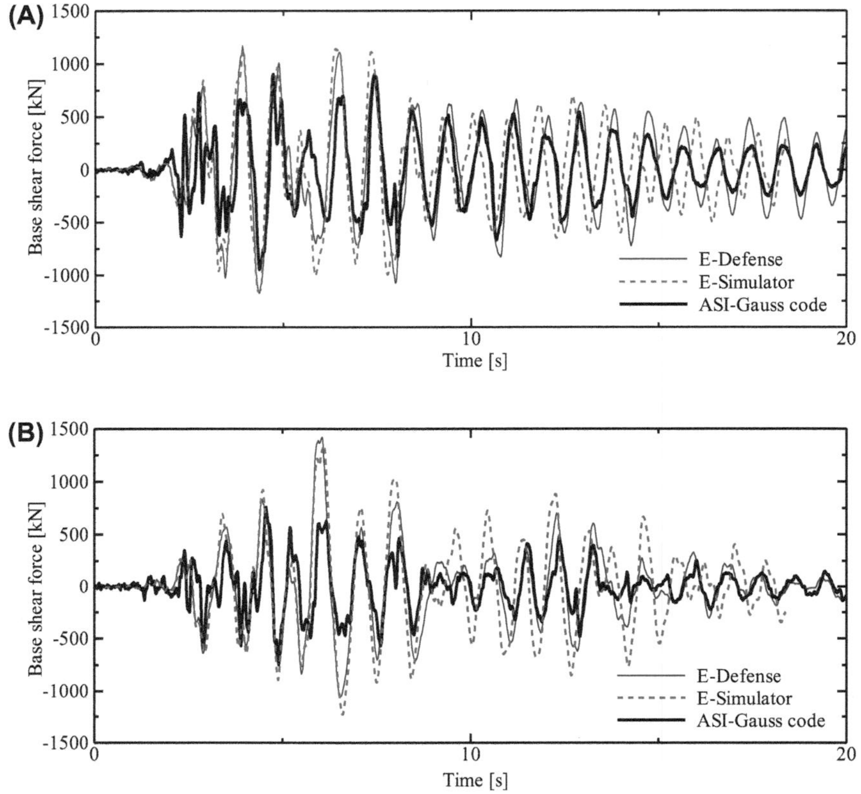

Figure 3.13 Comparison of the time histories of the shear force on the first floor. (A) *X*-direction and (B) *Y*-direction.

The interstory drift angles were calculated from the lateral displacements at the center of the second floor. The shear forces were calculated as the summation of the concentrated mass multiplied by the acceleration at the center of gravity of each floor. The experimental result obtained by the E-Defense is also plotted in the figures. The interstory drift angle along the *X*-axis obtained by the E-Simulator, as shown in Fig. 3.12A, agrees well with the experiment up to approximately 7 s. However, the differences in the phase and the amplitude become larger after 7 s, where the amplitude of the input ground motion decreases. The results obtained by the ASI-Gauss code, on the other hand, show good agreement with the experimental results in both the phase and the amplitude. The interstory drift angle along the *Y*-axis obtained by the E-Simulator, as shown in Fig. 3.12B, shows a good resemblance to the peak values near 6 s, but the angle drifts constantly in the positive direction in the later part of the experiment. Although the peak values near 6 s show smaller values in the result obtained by the ASI-Gauss code, the behaviors in the other parts are well documented. The time histories of the shear force shown in Fig. 3.13 seem to show similar tendencies, as in Fig. 3.12, where the peak values are achieved by the E-Simulator and the behaviors as a whole seem to be well recognized by the ASI-Gauss code.

Figs. 3.14 and 3.15 show the deformations and the distributions of the plastic region obtained by both analyses at 6.0 s. The deformation is magnified 10 times, and the colors represent the distribution of the equivalent stresses in Fig. 3.14 and the distribution of the yield function values f_y, defined in Eq. (3.4), in Fig. 3.15. The f_y values in the elements shown in Fig. 3.15 are linearly interpolated using the values at the both ends of the elements. Fig. 3.14 shows that the plastic regions around connections are localized and the positions of the plastic regions are the same as the positions of the plastic hinges (indicated with red color [gray in print version]) that are shown in Fig. 3.15. Therefore, the use of the plastic hinge model seems to be a good approximation of the plastic behavior obtained by the E-Simulator model. A torsional deformation can be clearly identified in the E-Simulator model, whereas less torsional deformation is observed in the ASI-Gauss model. The difference between the behaviors likely results

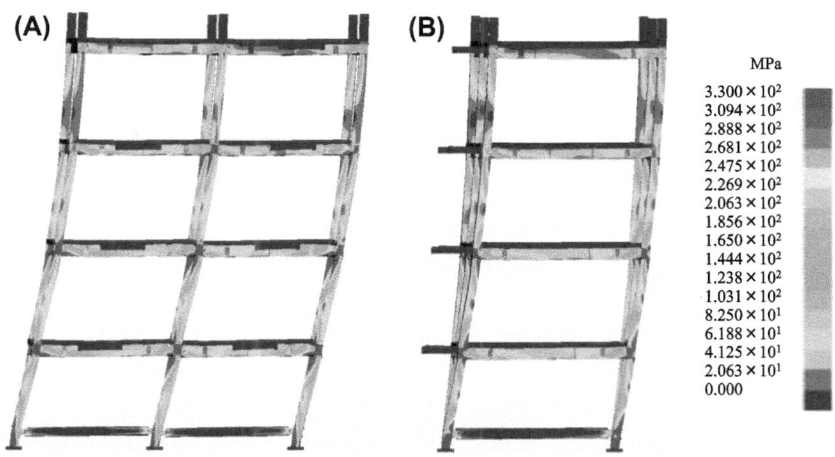

Figure 3.14 Deformation (magnified 10 times) and distribution of the equivalent stress at 6.0 s obtained by the E-Simulator. (A) *X*-direction and (B) *Y*-direction.

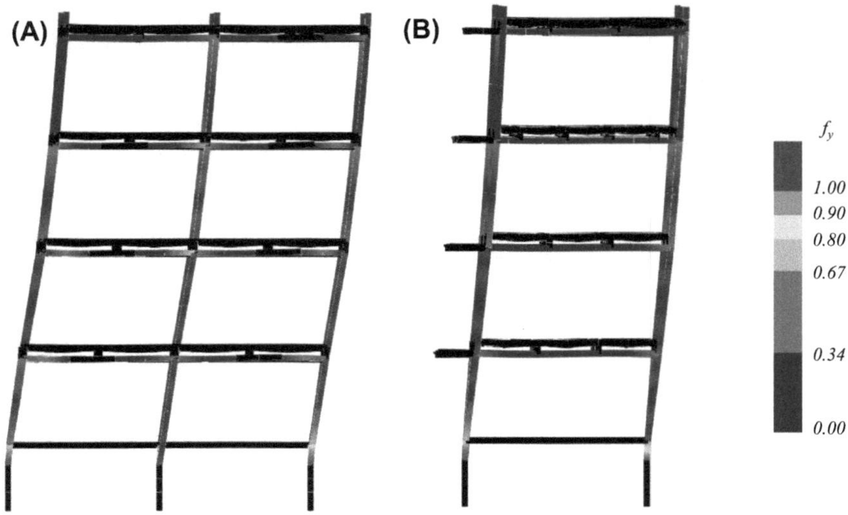

Figure 3.15 Deformation (magnified 10 times) and distribution of the yield function value at 6.0 s obtained by the ASI-Gauss code. (A) *X*-direction and (B) *Y*-direction.

from a lack of structural strength in the exterior walls in the ASI-Gauss model because an E-Simulator model without exterior walls (Miyamura et al., 2011) also showed less torsional deformation. However, the torsional deformation was not clearly observed in the full-scale experiment. Apart from these results, the advantages of the ASI-Gauss code, i.e., requiring short computational time and few memory resources, should be well noted. The disadvantage of the code, on the other hand, is the difficulty of modeling the structure in detail. For example, it cannot express local buckling in the frames.

3.4 SUMMARY

The ASI-Gauss technique, an improved version of the ASI technique, is described and some code V&V results are shown in this chapter. They indicate that the modified technique can supply highly accurate elasto-plastic solutions with minimum subdivision of finite elements and can simulate actual behaviors of frame models with short computational time and few memory resources. Consideration of member fracture, which is another noteworthy characteristic of the ASI-Gauss technique, is described in the next chapter along with some validation results.

REFERENCES

Akiba, H., et al., 2006. Large scale drop impact analysis of mobile phone using ADVC on Blue Gene/L. In: Proceedings of the International Conference on High Performance Computing Networking and Storage (SC06). Tampa, USA.

Architectural Institute of Japan, 1990. Recommendations for the Design and Fabrication of Tubular Structures in Steel. Architectural Institute of Japan (in Japanese).

Isobe, D., Han, W.S., Miyamura, T., 2013. Verification and validation of a seismic response analysis code for framed structures using the ASI-Gauss technique. Earthquake Engineering and Structural Dynamics 42 (12), 1767–1784.

Lynn, K.M., Isobe, D., 2007. Finite element code for impact collapse problems of framed structures. International Journal for Numerical Methods in Engineering 69 (12), 2538–2563.

Miyamura, T., Ohsaki, M., Kohiyama, M., Isobe, D., Onda, K., Akiba, H., Hori, M., Kajiwara, K., Ine, T., 2011. Large-scale FE analysis of steel building frames using E-Simulator. Progress in Nuclear Science and Technology 2, 651–656.

Toi, Y., 1991. Shifted integration technique in one-dimensional plastic collapse analysis using linear and cubic finite elements. International Journal for Numerical Methods in Engineering 31, 1537–1552.

Toi, Y., Isobe, D., 1993. Adaptively shifted integration technique for finite element collapse analysis of framed structures. International Journal for Numerical Methods in Engineering 36, 2323–2339.

Website of ASEBI at E-Defense, NIED. Project E-Defense Tests on Full-Scale Four-Story Steel Building. https://www.edgrid.jp/.

Website of Hyogo Earthquake Engineering Research Center (E-Defense), National Research Institute of Earth Science and Disaster Prevention (NIED), Japan, E-Defense HP. http://www.bosai.go.jp/hyogo/research/project_result/pdf/steel_3.pdf.

Yamashita, T., Miyamura, T., Akiba, H., Kajiwara, K., 2012. Verification of finite element elastic-plastic buckling analysis of square steel tube column using solid element. In: Transactions of the Japan Society for Computational Engineering and Science. Paper No. 20120019, (in Japanese).

Yoshimura, S., Shioya, R., Noguchi, H., Miyamura, T., 2002. Advanced general-purpose computational mechanics system for large scale analysis and design. Journal of Computational and Applied Mathematics 149, 279–296.

CHAPTER FOUR

Member-Fracture, Contact, and Contact-Release Algorithms

4.1 INTRODUCTION

Considerations of member fracture, elemental contact, and contact release are inevitable to simulate progressive collapse behaviors of structures. However, a precise and detailed consideration of these phenomena may lead to excessive consumption of memory resources and CPU time. In this chapter, some algorithms simple enough to match the rough approximations of structural members by beam elements are described.

4.2 MEMBER-FRACTURE ALGORITHM

Fig. 4.1 shows the locations of the numerical integration points for each stage of the ASI-Gauss technique (Lynn and Isobe, 2007a). The numerical integration points in the elastic stage are placed at the locations mentioned in the previous chapter, as shown in Fig. 4.1A. Once the yielding is determined in an element, the plastic hinge is expressed by shifting the numerical integration point to the opposite end of the fully plastic section, according to Eq. (3.1). For example, if a fully plastic section initially occurs at the upper end of an element ($r = r_1 = 1$), as shown in Fig. 4.1B, then the numerical

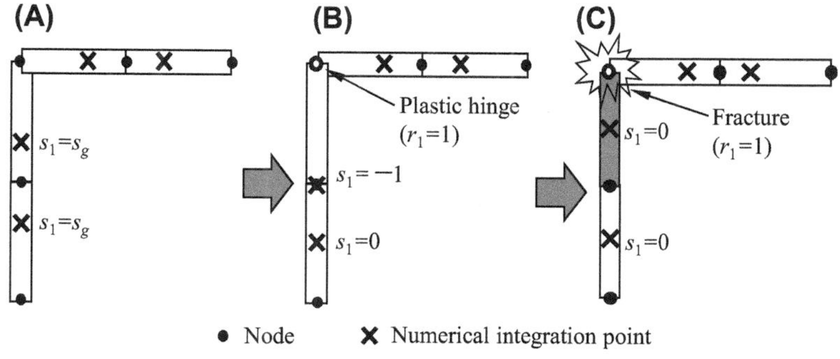

Figure 4.1 Locations of numerical integration points for each stage of the ASI-Gauss technique. (A) Elastic stage, (B) plastic stage, and (C) fracture stage.

Progressive Collapse Analysis of Structures
ISBN: 978-0-12-812975-3
http://dx.doi.org/10.1016/B978-0-12-812975-3.00004-9

integration point is shifted immediately to the lower end of the element ($s = s_1 = -1$). At the same time, the numerical integration point of the adjacent element that forms the same member is shifted back to its midpoint ($s = s_1 = 0$), where it is optimal for one-point integration in the Gaussian quadrature.

A member fracture is basically determined only at the fully plastic section. The fracture conditions of the member joints are considered by examining the curvatures, axial tensile strain, and shear strains in an element as follows:

$$\left|\frac{\kappa_x}{\kappa_{x0}}\right| - 1 \geq 0 \quad \text{or} \quad \left|\frac{\kappa_y}{\kappa_{y0}}\right| - 1 \geq 0 \quad \text{or} \quad \left|\frac{\gamma_{xz}}{\gamma_{xz0}}\right| - 1 \geq 0 \quad \text{or}$$
$$\left|\frac{\gamma_{yz}}{\gamma_{yz0}}\right| - 1 \geq 0 \quad \text{or} \quad \left(\frac{\varepsilon_z}{\varepsilon_{z0}}\right) - 1 \geq 0 \tag{4.1}$$

where κ_x and κ_y, γ_{xz} and γ_{yz}, ε_z, and κ_{x0}, κ_{y0}, γ_{xz0}, γ_{yz0} and ε_{z0} are the curvatures around x- and y-axes, the shear strains along x- and y-axes, the axial tensile strain and the critical values for these strains. One can fix the critical strain values to those values actually obtained from experiments concerning the fracture of joint bolts, such as those in Hirashima et al. (2007).

Once a fractured section is determined on either end of the element, it is expressed by reducing the sectional forces of the element immediately after the occurrence of the fractured section, as shown in Fig. 4.1C. The element, therefore, does not behave as a structural component after its section has fractured. The released force vector $\{F\}$, which operates on the element at the next step if the fractured section first occurs at the upper end of an element ($r = r_1 = 1$), is expressed by the following equation:

$$\{F\} = {}^{n}l[B(-1)]^T\{\sigma(1)\} \tag{4.2}$$

Similarly, if a fully plastic section or a fractured section first occurs at the lower end of the element ($r = r_1 = -1$), the numerical integration point is shifted to the upper end of the element ($s = s_1 = 1$) and the released force vector is expressed by the following equation:

$$\{F\} = {}^{n}l[B(1)]^T\{\sigma(-1)\} \tag{4.3}$$

In the FEM, the mass of an element is distributed at its nodes. Therefore, the mass matrix needs to be recalculated according to the new connecting conditions of the elements when member fracture occurred. In the ASI-Gauss code, half of the nodal mass is redistributed at each separated node after the occurrence of fracture. It should be noted that fractured elements are still treated as a part of the continuous model in the calculation process; however, new virtual nodes for the fractured sections are added in the post-processing stage, and the elements with the virtual nodes are visualized as rigid bars thereafter.

4.3 ELEMENTAL-CONTACT ALGORITHM

A set of two elements in potential contact is first selected by evaluating the shortest distance between them using the geometric relations of the four nodes (Fig. 4.2A). By examining the angles between the assumed contact point M and two nodes, a final determination of contact is made if the angles $\angle MB_1B_2$ and $\angle MB_2B_1$ are both acute (Fig. 4.2B); the two elements will not come in contact if one of the angles is obtuse (Fig. 4.2C). Once they are determined to be in contact, the elements are bound with a total of four gap elements between the nodes (Fig. 4.2D). The gap elements are assumed to have physical and material properties, which are similar to the elements in contact. The sectional forces will be delivered through these gap elements to the connecting elements.

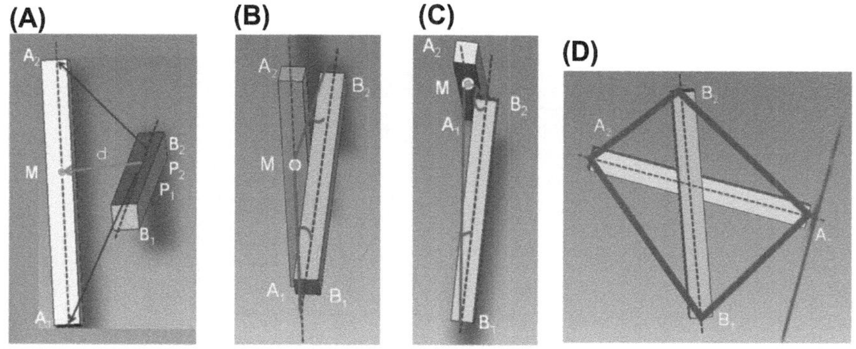

Figure 4.2 Elemental-contact algorithm. (A) Searching for shortest distance, (B) contact if both angles are acute, (C) pass by if one of both angles is obtuse, and (D) four gap elements connecting two elements in contact.

4.4 CONTACT-RELEASE ALGORITHM

After a certain time from contact, the four gap elements are removed from the numerical model to simulate the rebound behavior, if any, of the elements that were in contact. The binding time of the gap elements should be the time during which the two elements involved in an impact stay physically in contact in actual situations. Therefore, it is automatically determined using time histories of the deformation of the four gap elements (Lynn and Isobe, 2007a). There are two options for the release condition; the gap elements are removed (1) if the value of equivalent displacement $\sqrt{u_x^2 + u_y^2 + u_z^2}$ of the four nodes in Fig. 4.2D is decreased to a certain ratio, or (2) if at least one of the mean values of displacements of the four nodes in three-dimensional directions, $|u_x|$, $|u_y|$, $|u_z|$, is decreased to a certain ratio. The ratio is defined here as a contact-release ratio and the concept of contact release is shown in Fig. 4.3. The release condition is selected depending on the contact mode, and the values of contact-release ratio are determined by some simple experimental tests; a value of 0.2 with condition (1) is used for vertical contact, and a value of 0.95 with condition (2) is used for horizontal contact in the numerical examples described in this book.

4.5 EVALUATION OF THE ALGORITHMS

Impact phenomenon of two beams was simulated to evaluate the creditability of the elemental-contact algorithm (Lynn and Isobe, 2007b). The analyzed model, its physical properties, and material constants are

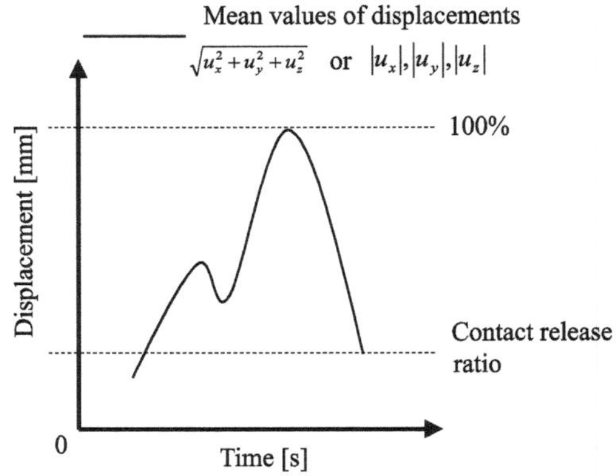

Figure 4.3 General concept of contact release.

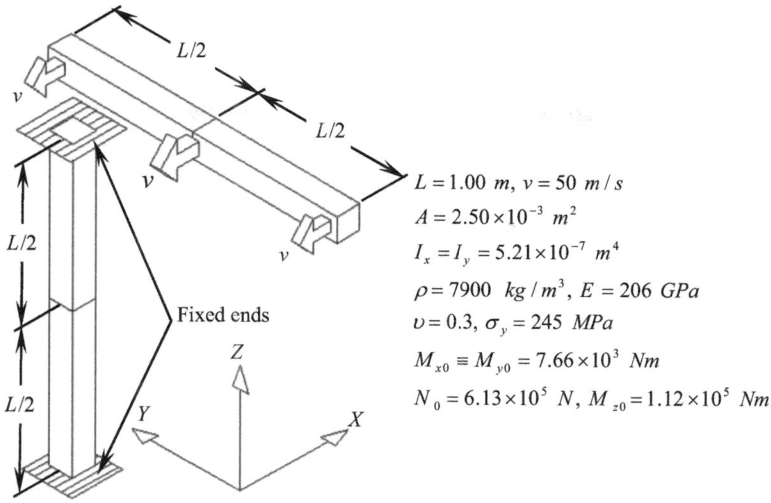

Figure 4.4 Analyzed two-beam model.

shown in Fig. 4.4. One of the beams was fixed at its both ends, while the other one moved freely. Initial velocity of 50 m/s was applied to the free beam. The beams were modeled with two elements per member. The ASI-Gauss technique was applied, and the updated Lagrangian formulation was used since large deformations were expected in the analysis. Damping effect was not considered in the analysis. The time increment was set to 10 μs and the calculation was done for a total of 3000 steps (0.03 s).

Fig. 4.5 shows the sum of kinetic and potential energies, strain energy and the total of all three for the whole model. Energies for both the colliding and the fixed beams constituted by a total of four elements are calculated by using the following equations:

$$\mathrm{KE} = \frac{1}{2}\sum_{i=1}^{4} \rho A \frac{l_i}{2}\left(v_{i1}^2 + v_{i2}^2\right) \tag{4.4a}$$

$$\mathrm{PE} = \sum_{i=1}^{4} \rho A \frac{l_i}{2} g(h_{i1} + h_{i2}) \tag{4.4b}$$

$$\mathrm{SE} = \frac{1}{2}\sum_{i=1}^{4} \{\sigma_i\}\{\varepsilon_i\} l_i \tag{4.4c}$$

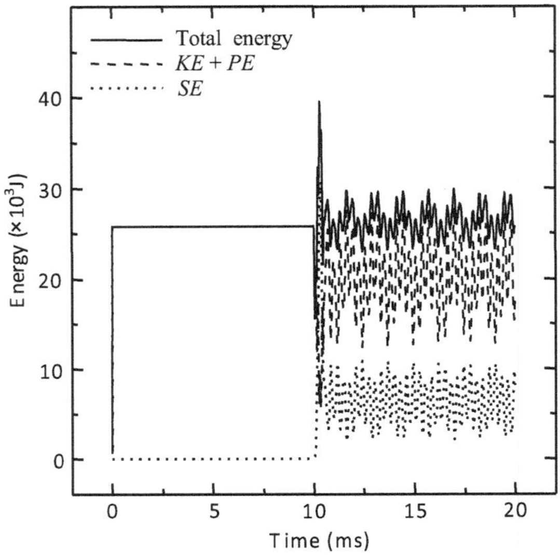

Figure 4.5 Kinetic, potential, and strain energies in the two-beam model.

where KE, PE, and SE stand for kinetic energy, potential energy, and strain energy, respectively. $\{\sigma_i\}$, $\{\varepsilon_i\}$, h_i, l_i, and v_i represent the generalized sectional force vector, the generalized strain vector, z-coordinate, the length and velocity of the element E_i $(i = 1, 2, 3, 4)$, respectively. Subscripts 1 and 2 denote the nodes of the corresponding element.

As seen from the figure, kinetic and potential energies occupy the entire total energy before the impact takes place. After the impact, both beams deflect and vibrate as seen in Fig. 4.6. As a consequence, the strain energy is increased in the beams and the sum of the other energies decreases. Oscillations seen after the impact is due to the occurrence of longitudinal and bending waves. However, the total energy remains almost unchanged, and the elemental-contact algorithm was considered to be sufficiently reliable.

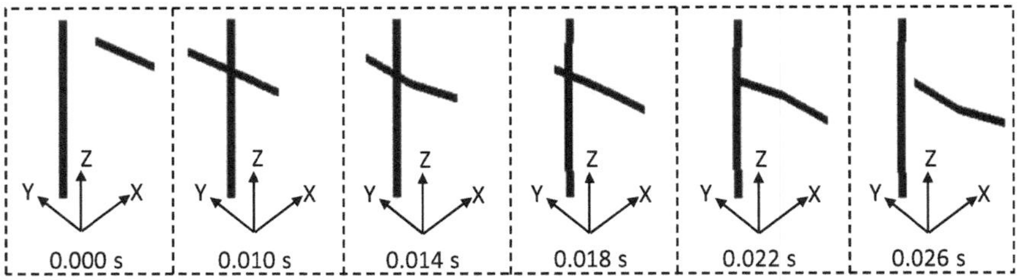

Figure 4.6 Impact analysis of a simple two-beam model.

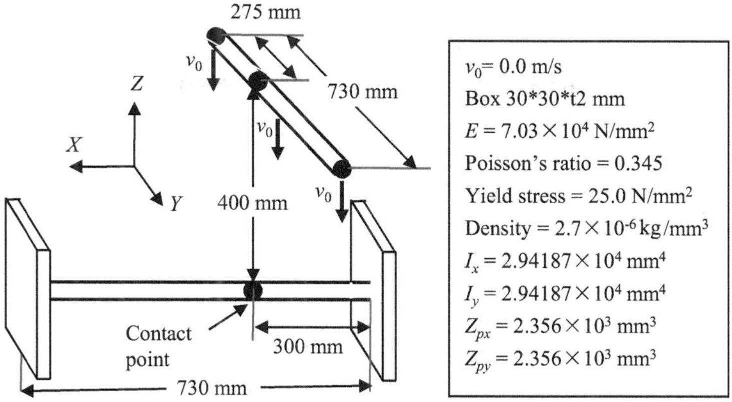

Figure 4.7 Experimental and numerical conditions for the impact test with two aluminum beams.

4.6 VALIDATION OF THE ALGORITHMS

To validate the algorithms mentioned earlier, a simple impact test of two beams was performed (Isobe, 2014). As shown in Fig. 4.7, one of the two aluminum beams was rigidly clamped at the both ends, whereas the other was silently dropped with an initial velocity of 0 m/s from the height

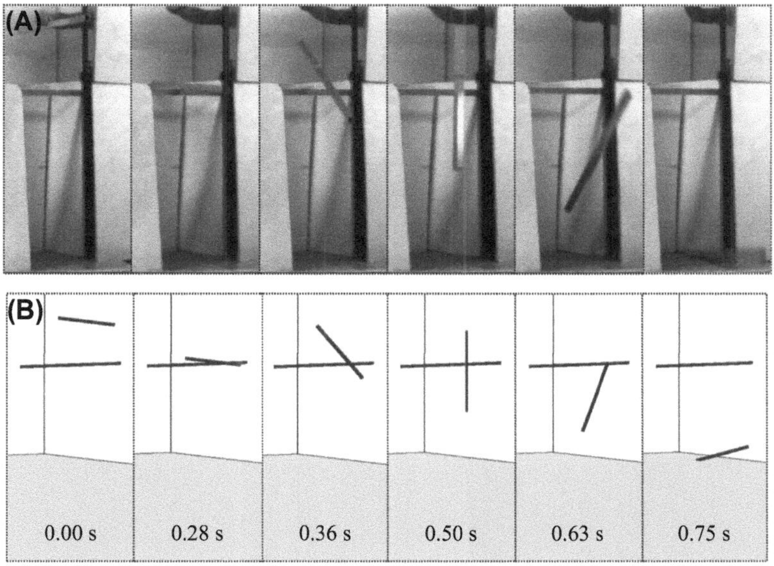

Figure 4.8 Impact and contact behaviors of two aluminum beams. (A) Experimental and (B) numerical results.

of 400 mm above the clamped beam. The planned contact point between the two beams is as indicated in the figure. The impact and contact phenomena were observed using a high-speed camera as shown in Fig. 4.8A and the numerical result was obtained by the ASI-Gauss code as shown in Fig. 4.8B. The two results agree very well, indicating that the contact algorithm implemented in the numerical code performed as intended to model the impact, contact, and rebound phenomena.

4.7 SUMMARY

The numerical algorithms required to simulate progressive collapse behaviors of structures, such as member fracture, elemental contact, and release, are described in this chapter. The simplicities of the algorithms match the simplified approximations of structural members using beam elements, while their reliabilities are well guaranteed according to the verification and validation results shown in this chapter. These algorithms are implemented to a numerical code using the ASI-Gauss technique, and applied to various collapse problems of structures. Some of the applications are shown in the following chapters.

REFERENCES

Hirashima, T., Hamada, N., Ozaki, F., Ave, T., Uesugi, H., 2007. Experimental study on shear deformation behavior of high strength bolts at elevated temperature. Journal of Structural and Construction Engineering AIJ 621, 175–180 (in Japanese).

Isobe, D., 2014. An analysis code and a planning tool based on a key element index for controlled explosive demolition. International Journal of High-Rise Buildings 3 (4), 243–254.

Lynn, K.M., Isobe, D., 2007a. Finite element code for impact collapse problems of framed structures. International Journal for Numerical Methods in Engineering 69 (12), 2538–2563.

Lynn, K.M., Isobe, D., 2007b. Structural collapse analysis of framed structures under impact loads using ASI-Gauss finite element method. International Journal of Impact Engineering 34 (9), 1500–1516.

CHAPTER FIVE

Aircraft-Impact Analysis of the World Trade Center Tower

5.1 INTRODUCTION

The terrorist attack on the World Trade Center (WTC) towers in New York on September 11, 2001 was an unprecedented tragedy in the history of architecture. The Twin Towers, WTC 1 and WTC 2, were engulfed in flames caused by jet fuel, and finally, both towers collapsed to the ground, causing thousands of people to be trapped in the rubble. Both towers collapsed at a very high speed nearly equal to a free fall. Official statements regarding the incident were released by the Federal Emergency Management Agency (FEMA) in 2002 (ASCE/FEMA, 2002) and the National Institute of Standards and Technology (NIST) in 2005 (NIST, 2005) and 2008 (NIST, 2008). In its report, FEMA concluded that the heat of burning jet fuel induced additional stresses into the damaged structural frames simultaneously softening and weakening these frames, and this additional loading and the resulting damage were sufficient to induce the collapse of the towers. Many detailed numerical analyses were performed by NIST, and they concluded that "the WTC towers likely would not have collapsed under the combined effects of aircraft impact damage and the extensive, multifloor fires, if the thermal insulation had not been widely dislodged or had been only minimally dislodged by aircraft impact." However, the specific cause of such a high-speed collapse still remains unresolved.

Regarding this unresolved structural problem, an aircraft-impact analysis of a full-scale WTC tower was performed to seek the possibility that the impact itself was the main cause of the high-speed collapse (Isobe et al., 2012). It is clear that the towers experienced an extreme dynamic load that no other high-rise buildings in history have ever experienced (Fig. 5.1). Full-scale models of WTC 2 and an aircraft were constructed, and an impact analysis was performed.

5.2 NUMERICAL MODEL AND CONDITIONS

An aircraft-impact analysis of a full-scale model of WTC 2 (the south tower) was performed using the numerical code introduced in

Progressive Collapse Analysis of Structures
ISBN: 978-0-12-812975-3
http://dx.doi.org/10.1016/B978-0-12-812975-3.00005-0

Figure 5.1 Imminent impact of UA-175 on WTC 2. ©*Associated Press/AFLO.*

Chapters 2–4. The details of the structural members and the construction data of the tower were extracted from the official reports (ASCE/FEMA, 2002; AIJ, 2003; NIST, 2005). The tower and the B767-200ER aircraft, which collided in the tower, were modeled with linear Timoshenko beam elements. All members were subdivided into two elements. Fig. 5.2 shows a schematic view of the WTC 2 tower model and the aircraft model. The tower model consisted of 604,780 elements, 435,117 nodes, and 2,608,686 degrees of freedom, and the aircraft model consisted of 4322 elements, 2970 nodes, and 17,820 degrees of freedom. Fig. 5.3 shows a cross-section of the tower. The floor was coarsely divided into two sections, the external and core structures, which had different structural roles. The external structure was primarily designed to sustain wind loads, whereas the main double trusses connected to the core structure supported the floor loads. The entire tower was subjected to a designed vertical load of 2890 MN and 40% of the engineered capacity load of 300 MN. The ratio between the loads of the core and external structures was adjusted to a value of

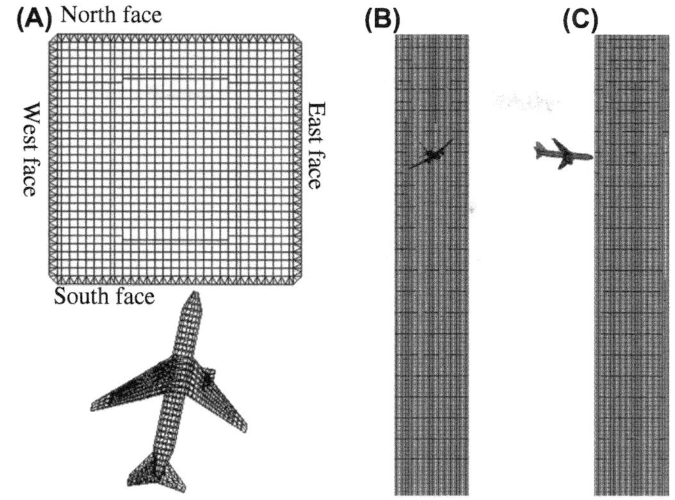

Figure 5.2 Numerical model of the WTC 2 and the aircraft. (A) Upper view, (B) south view, and (C) east view.

approximately 6:4 (ASCE/FEMA, 2002). The structural members of the B767-200ER aircraft were assumed to be box-section type beams with the material properties of extra-super duralumin. The total mass of the aircraft at the time of impact was assumed to be 142.5 tons, which is the sum of the mass of the aircraft (112.5 tons) and the jet fuel (30 tons). The mass of each engine was set to 19.315 tons. The posture of the aircraft

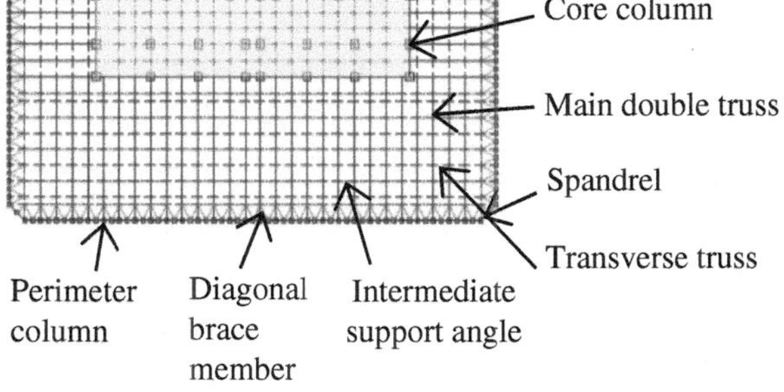

Figure 5.3 Cross-section of the WTC tower.

at the time of impact was fixed, by referring to the reports, as follows: the nose of the aircraft tilted 11.5 degrees to the east and 5 degrees downward, and the left wing inclined 35 degrees downward. The aircraft was set to collide into the 81st floor on the south face of the WTC 2 at a speed of 590 mph (262 m/s).

The yield condition used in this chapter is expressed as follows:

$$f = \left(\frac{M_x}{C_M M_{x0}}\right)^2 + \left(\frac{M_y}{C_M M_{y0}}\right)^2 + \left(\frac{N}{N_0}\right)^2 - 1 = f_y - 1 = 0 \qquad (5.1)$$

where C_M is the member joint strength ratio, which indicates the moment capacity ratio of the member connection to the adjacent member; it is used to explicitly express the influence of the connection strength. The values of the ratio were fixed on the basis that the plastic moment capacity of some of the column connections in the WTC towers was weak compared to those of their members (ASCE/FEMA, 2002). Most of the column connections were weak compared to fully rigid connections, though they fully satisfied the building codes of New York City. For example, the simple moment capacity of the bolt group in the perimeter columns was 20%–30% of the plastic moment capacity of the column member itself (ASCE/FEMA, 2002). The beam-to-column connections between the main double trusses and the core columns had simple connections to sustain the static, vertical loads, which satisfied the initial structural design. However, the member joint strength ratio was set to a value of 1.0 in this analysis for simplicity, and the effects of the connection strength were evaluated in the progressive collapse analyses that are shown in Chapter 6. An elasto-plastic body with a strain hardening of 1/100th of the elastic modulus was assumed for all structural members, and the effect of the strain rate on the yield strength was considered by applying the Cowper–Symonds equation (Jones, 1989) as follows:

$$\frac{\sigma'}{\sigma_0} = 1 + \left(\frac{\dot{\varepsilon}}{D}\right)^{\frac{1}{q}} \qquad (5.2)$$

where σ', σ_0, and $\dot{\varepsilon}$ are the dynamic yield stress, static yield stress, and strain rate, respectively, and D and q are the material constants. The values of 40.4 and 5 were used for D and q of steel, and 6500 and 4 were used for duralumin.

In this analysis, the critical curvature and the axial tensile strain values for member fracture were determined by repeatedly comparing the impact damage and the motion of the engines with the observed data. As mentioned in Chapter 2, the implicit scheme with a distributed mass matrix was adopted, and numerical damping was applied; the integration parameters δ and β were set to 5/6 and 4/9, respectively, for the Newmark's β method (Press et al., 1992). An updated Lagrangian formulation was used because large deformations were expected in the analysis, and the conjugate gradient method was used as the solver to reduce computational cost. The time increment, Δt, was set to 0.2 ms, and the numerical code was executed on a high-performance computer (1.4 GHz Itanium*2, 8GB RAM) for approximately 2 months to obtain the results for an actual crash duration of 0.7 s.

5.3 NUMERICAL RESULTS

Motion views of the aircraft during impact are shown in Figs. 5.4 and 5.5, in which the fractured elements are omitted. The left engine rapidly reduces its speed as it enters the core structure; however, the right engine glances off the core structure and passes through the northeast corner of the tower. Table 5.1 shows the impact timeline of the nose and both engines of the aircraft. The velocity curve of the right engine is shown in Fig. 5.6 along with its observational data (ASCE/FEMA, 2002). The letters A to F indicated in the figure represent the corresponding phases listed in Table 5.1.

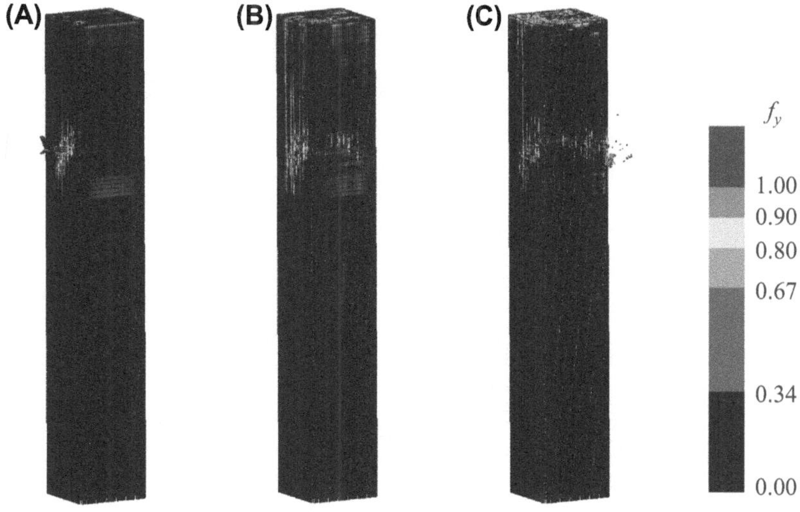

Figure 5.4 Collision of the aircraft to the WTC 2 (global view). (A) 0.12 s after impact, (B) 0.28 s after impact, and (C) 0.56 s after impact.

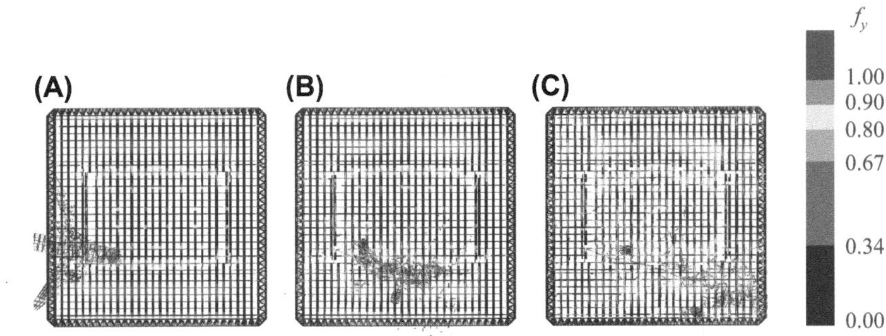

Figure 5.5 Motion of the fuselage and engines (upper view). (A) 0.12 s after impact, (B) 0.28 s after impact, and (C) 0.56 s after impact.

A comparison between the numerical results and the observational data (NIST, 2005) for the damaged surfaces is shown in Fig. 5.7. Both results are in good agreement with the observational data. Fig. 5.8 depicts the core column numbers in the WTC tower and their relative locations to the point of aircraft impact. Fig. 5.9 shows the transition of the axial forces in two of the core columns (No. 508 and No. 1001) at increments of 10 stories from the ground floor to the top floor. The letters A to F indicated in the figure, again, represent the corresponding phases given in Table 5.1. Core column No. 1001 was compressed continuously until the nose of the

Table 5.1 Impact Timelines of the Nose and Both Engines of the Aircraft

Phase	Elapsed Time (s)	Left Engine	Nose	Right Engine
		B767-200ER Aircraft		
	0.000	Nose hits outer-perimeter wall		
A	0.048		Hits core structure	
	0.072	Hits outer-perimeter wall		
B	0.076			Hits outer-perimeter wall
C	0.128	Hits core structure		
D	0.168	Hits core column No.902		
E	0.504	Hits core column No.903		
F	0.512			Hits east side outer-perimeter wall

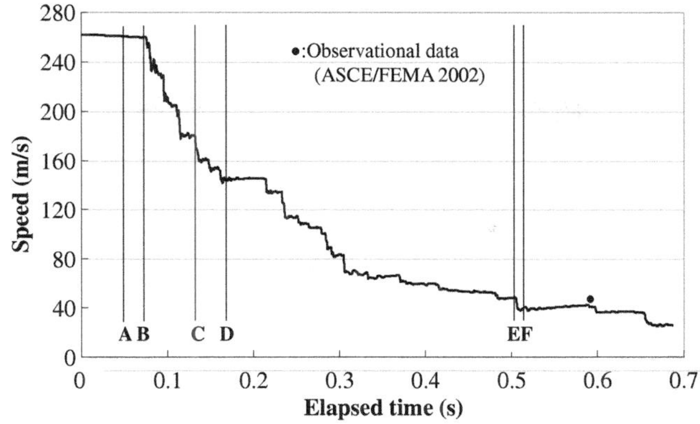

Figure 5.6 Velocity curve of the right engine.

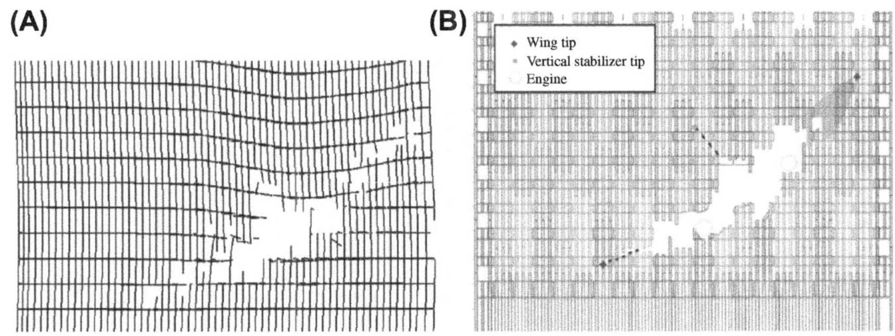

Figure 5.7 Damage to the south face of the WTC 2. (A) Numerical result and (B) observational data (NIST, 2005).

aircraft reached to the core structure (Phase A in Fig. 5.9A), and a shock wave due to this impact propagated in the horizontal and vertical directions thereafter. The wave reached core column No. 508 after a slight delay (see Fig. 5.9B). The compression that occurred in column No. 1001 decreased instantaneously at the floors above the impact point. At the lower levels, the compression transitioned into tension immediately after the left engine hit the core structure (Phase C), and the axial forces continued to vibrate with a large amplitude. Note that the lower the height, the larger the amplitude becomes. For example, as shown in Fig. 5.10, column No. 1001 at the 60th floor moved vertically for 25 cm in 0.2 s as a result of this dynamic unloading. The beam-to-column connections were likely to have failed in this process, but nevertheless, the floors could not have followed this column movement due to their large momentum.

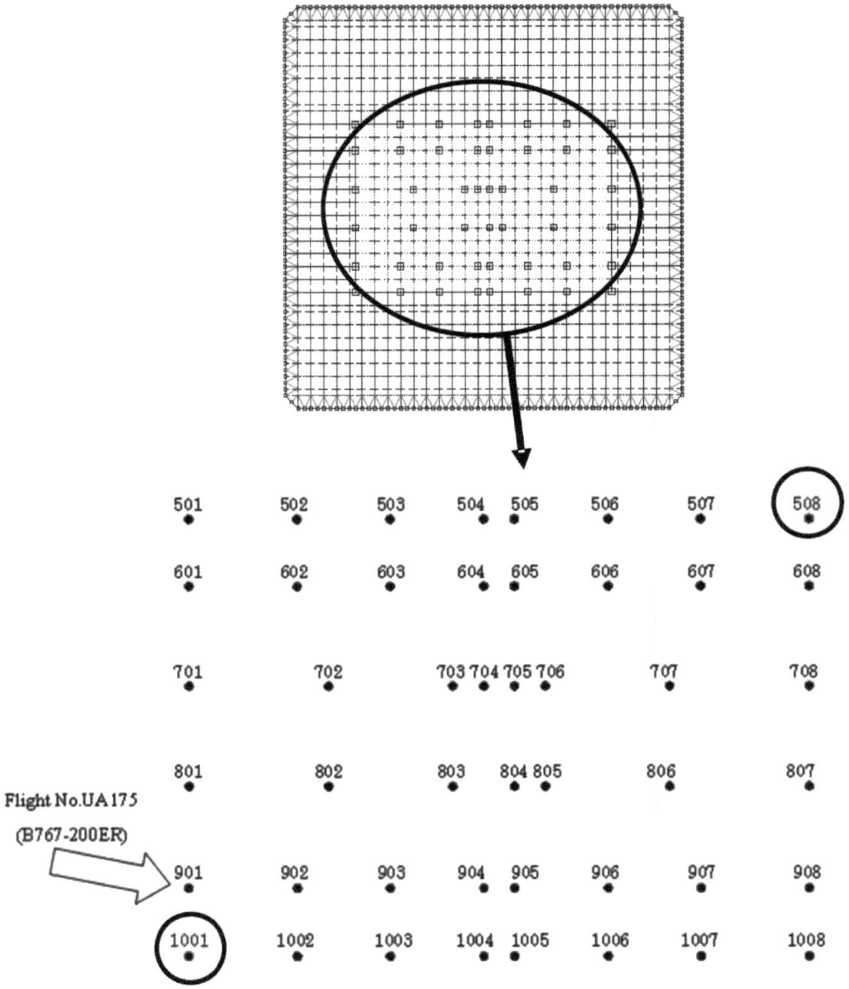

Figure 5.8 Location of the core columns.

The temporal change in the distribution of the fractured main double trusses is extracted, as shown in Fig. 5.11, to investigate the cause of the dynamic unloading. The fractured joints in the main double trusses are indicated as black dots in this figure. The propagation of fractured joints from the impact point to the upper and lower levels is apparent. The propagation speed agreed well with the theoretical speed at which the longitudinal wave travels through the columns. The shock wave due to the impact propagated through the tower and was transmitted to the other columns at the ground

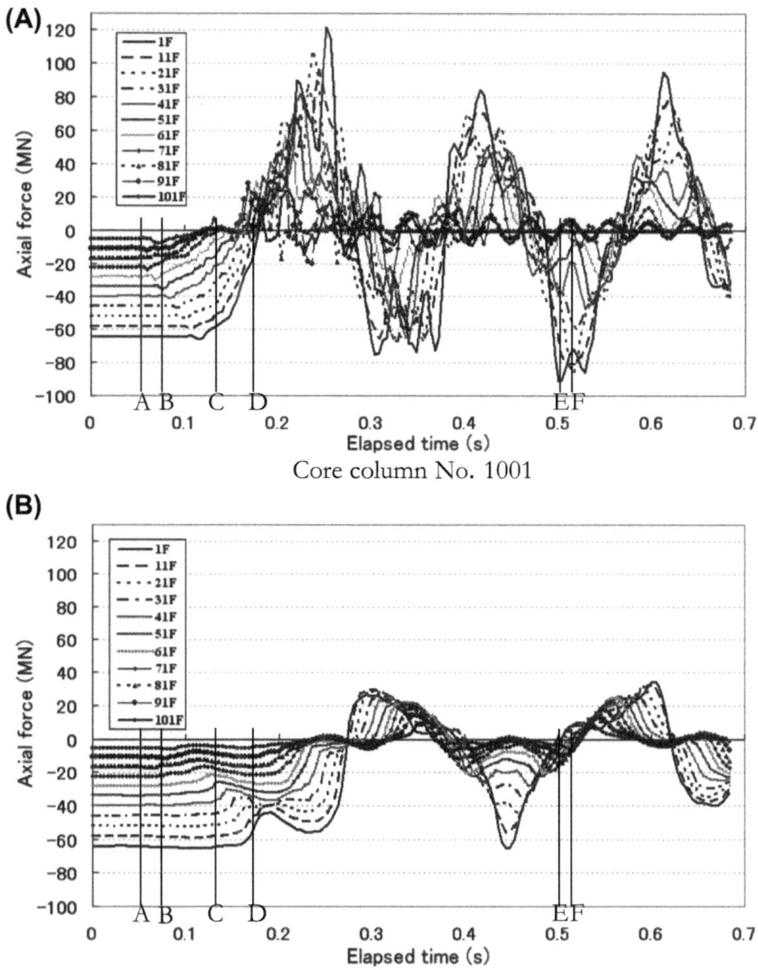

Core column No. 508

Figure 5.9 Transition of the axial forces occurred in the core columns. (A) Core column No. 1001 and (B) core column No. 508.

level or top level of the tower, which destroyed each member connection as the wave moved along the columns.

The assumed scenario regarding the total collapse of the towers is described as follows: the impact of the aircraft destroyed several connections between the core columns and the main double trusses that supported the floor slabs, which resulted in a middle-class unloading of the core columns. Deformation due to the unloading caused other trusses to disconnect, which

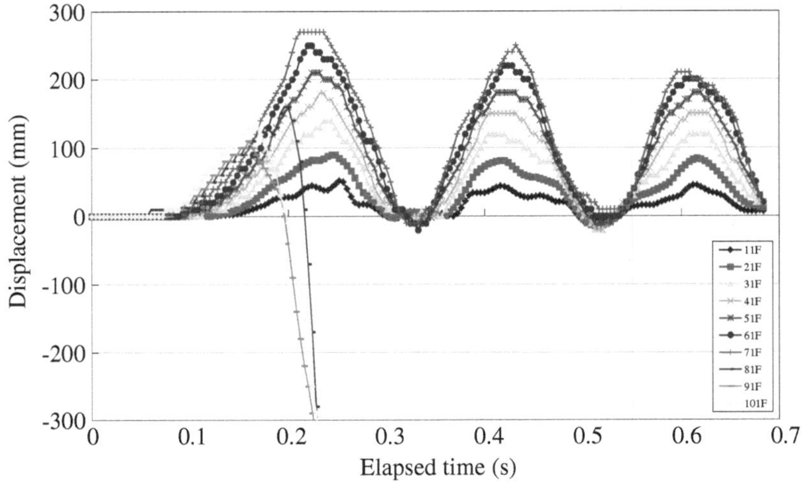

Figure 5.10 Displacements occurred in vertical direction.

relieved additional loads from the core columns. This disconnect-and-release process advanced in a very short period of time as a chain reaction. This dynamic unloading phenomenon is called a spring-back phenomenon, because it might have occurred in many columns like a compressed spring being quickly released. This process was substantial enough to produce a large tensile force in the columns, which may have triggered the total

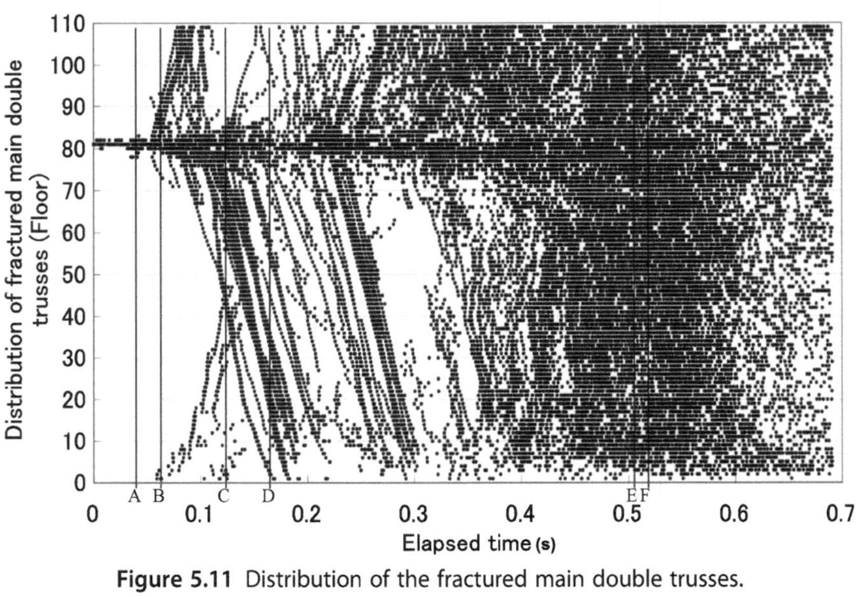

Figure 5.11 Distribution of the fractured main double trusses.

fracture of some column connections and made the columns similar to toy blocks (with no connections) piled on top of one another. The damage to the lower levels was greater because the tensile force increased along with a decrease in height, as observed in Fig. 5.9. The compression strength of the columns remained unchanged, however, which prevented the tower from falling down for some time.

5.4 SUMMARY

The numerical results of the aircraft-impact analysis revealed the phenomena that might have taken place in the WTC towers; how the main wings of the aircraft cut through the perimeter columns and spandrels, how the debris moved through the building, and where it caused damage. Moreover, the velocity reduction of the right engine and the location from where it moved out of the building were in good agreement with the observed data. The numerical results also showed instant disconnections between the core columns and the main double trusses during the impact, and spring-back phenomena of the core columns due to very rapid unloading. The spring-back phenomena were fatal enough to produce a gigantic tensile force on the columns, which might have triggered the total fracture of some connections. Chapter 6 will show some fire-induced collapse analyses of high-rise towers conducted to simulate the event which took place after the collision of the aircraft.

REFERENCES

Architectural Institute of Japan (AIJ), 2003. Report on WTC Collapse Investigation. Special Investigation Committee. Architectural Institute of Japan (AIJ) (in Japanese).

ASCE/FEMA, 2002. World Trade Center Building Performance Study: Data Collection, Preliminary Observation and Recommendations. ASCE/FEMA.

Isobe, D., Thanh, L.T.T., Sasaki, Z., 2012. Numerical simulations on the collapse behaviors of high-rise towers. International Journal of Protective Structures 3 (1), 1−19.

Jones, N., 1989. Structural Impact. Cambridge University Press, New York.

NIST NCSTAR 1, 2005. Federal Building and Fire Safety Investigation of the World Trade Center Disaster: Final Report on the Collapse of the World Trade Center Towers. National Institute of Standards and Technology (NIST).

NIST NCSTAR 1A, 2008. Final Report on the Collapse of World Trade Center Building 7. National Institute of Standards and Technology (NIST).

Press, W.H., Teukolsky, S.A., Vetterling, W.T., Flannery, B.P., 1992. Numerical Recipes in FORTRAN: The Art of Scientific Computing. Cambridge University Press, New York.

CHAPTER SIX

Fire-induced Progressive Collapse Analysis of High-rise Buildings

6.1 INTRODUCTION

This chapter performs a fire-induced collapse analyses of high-rise building model without any initial structural damage to qualitatively verify the influences of fire patterns and structural parameters, such as the axial force ratio, the member strength of the outrigger truss system, and the connection strength, with respect to the collapse initiation time, i.e., the duration from the beginning of the fire until the beginning of the collapse (Isobe et al., 2012). These analyses were not performed on a full-scale model of the World Trade Center (WTC) tower because the qualitative information on the collapse behavior of high-rise buildings on fire and the contributions of their structural parameters to their collapse behaviors are, for some reason, the most desired information in terms of understanding the collapse phenomena associated with high-rise buildings. The information that is obtained should be applicable to all high-rise buildings, not just the WTC towers.

Although the behavior of a structure prior to the initiation of its collapse is nearly static, collapse behavior of a structure is a dynamic phenomenon and there are greater calculation requirements for the dynamic analysis of the entire sequence. Therefore, the ASI-Gauss code considering thermal deformations was efficiently applied, to the fire-induced collapse simulations of a high-rise building model with an outrigger truss system on the roof, to repress the calculation cost.

6.2 NUMERICAL MODEL AND CONDITIONS

Fire-induced collapse analyses were performed on a 30-story, 7-span building, as shown in Fig. 6.1. An outrigger truss system was placed on the roof of the model, and the influence of this system on the structural vulnerability of the building was verified. In addition, several structural parameters were varied to determine their contributions to the collapse behavior. The axial force ratio of the columns to the structural strength capacity was varied from 0.1

Progressive Collapse Analysis of Structures
ISBN: 978-0-12-812975-3
http://dx.doi.org/10.1016/B978-0-12-812975-3.00006-2

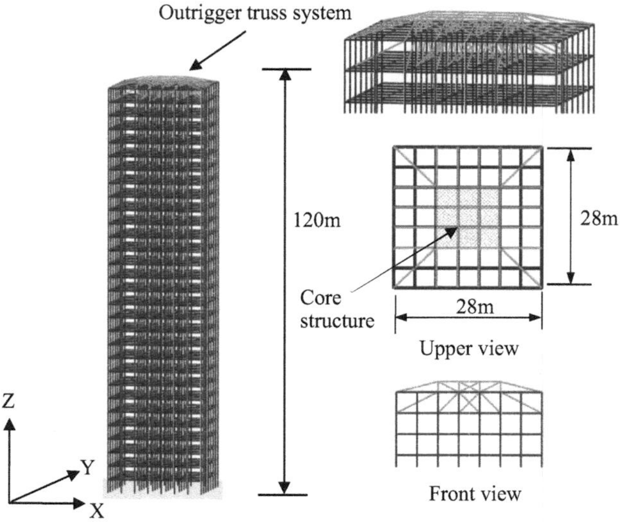

Figure 6.1 30-story 7-span building model.

to 0.5, which was altered by varying only the floor loads. The member joint strength ratio of every member connection in the model, C_M, was varied from 0.1 to 0.6, and the member strength ratio of the outrigger trusses to the strength of the beams on the first floor was varied from 0.0 (no outrigger truss system in place) to 2.0. The building model consisted of 9360 linear Timoshenko beam elements, 6644 nodes, and 39,600 degrees of freedom. The material properties including the critical fracture strains obtained from experiments performed on high-strength joint bolts (Hirashima et al., 2007) and the sectional parameters of the columns and beams are shown in Tables 6.1 and 6.2. Three simple fire patterns, as shown in Fig. 6.2, were applied to determine how the catenary action of the outrigger

Table 6.1 Material Properties of the Structural Members

Young's modulus (SN490B)	206 GPa
Yield strength (SN490B)	325 MPa
Poisson's ratio (SN490B)	0.3
Density (SN490B)	7.9×10^{-6} kg/mm^3
Critical bending strain for fracture[a]	3.333×10^{-4}
Critical shear strain for fracture[a]	1.300×10^{-2}
Critical axial tensile strain for fracture	0.17

[a]From Hirashima, T., Hamada, N., Ozaki, F., Ave, T., Uesugi, H., 2007. Experimental study on shear deformation behavior of high strength bolts at elevated temperature. Journal of Structural and Construction Engineering AIJ 621, 175–180 (in Japanese).

Table 6.2 Sectional Properties of the Structural Members

Floor	Cross Section
Cross Section of Columns	
1–5	Box-700 × 28
6–10	Box-600 × 25
11–15	Box-500 × 24
16–20	Box-400 × 23
21–30	Box-350 × 20
Cross Section of Beams	
1–10	H-700 × 300 × 13 × 24
11–20	H-700 × 250 × 12 × 22
21–30	H-700 × 200 × 12 × 22

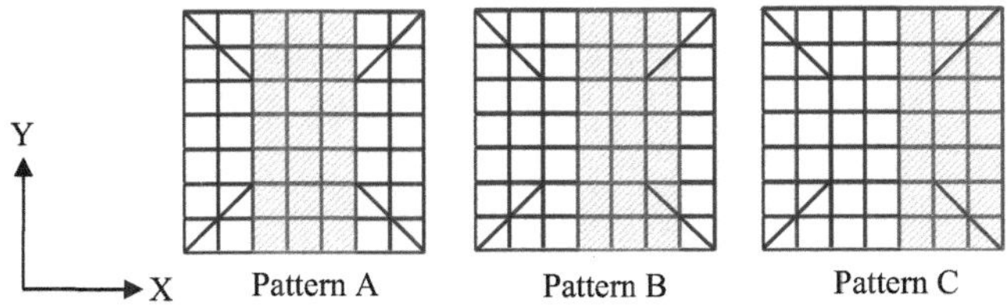

Figure 6.2 Assumed fire patterns (21–24F).

truss system contributed to gain time before collapse. These fire patterns comprised three adjacent rows of sections that were on fire, from the 21st to the 24th floors, located from the middle to the sides of the building. The height of the area on fire was assumed with regards to the WTC 2 case; the upper one-third to one-fourth of a section was treated as on fire. The structural damage due to the aircraft impact, as explained in the previous chapter, was neglected in this model; however, the reduction of the elastic modulus and yield strength of raw steel that occur with an increase in temperature (NIST, 2005) were incorporated as shown in Fig. 6.3. It was found that at 700°C, a typical fire temperature, the elastic modulus of steel decreases to 60% of its original value, and the yield strength is reduced to 10% of its original strength. To reduce complexity, a simple assumption was made that the temperature of the beam elements in the areas shown in Fig. 6.2 increased to 700°C in 7 min, as shown in Fig. 6.4. The temporal change in temperature follows a curve used in standard Japanese fire resistance tests. The model was dynamically analyzed throughout the entire calculation; however, the time increment was controlled to enable

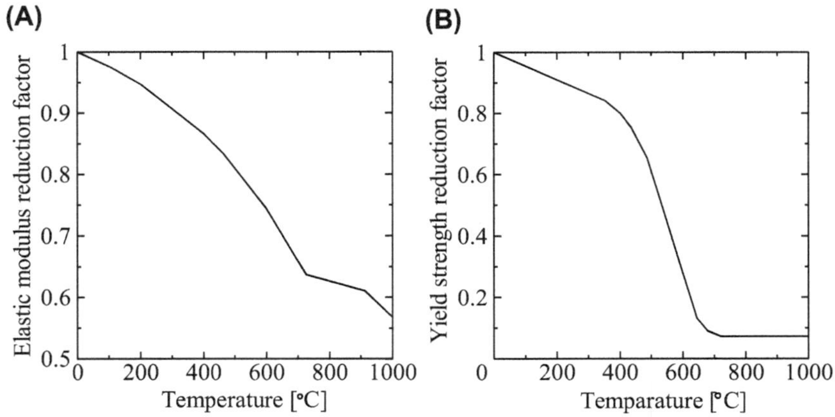

Figure 6.3 Strength reduction of steel with increasing temperature. (A) Elastic modulus and (B) yield strength.

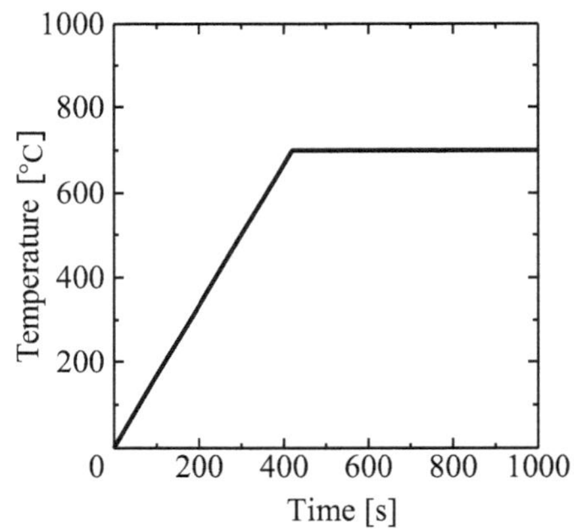

Figure 6.4 Assumed temporal change in temperature.

a continuous calculation from static to dynamic phenomena. Specifically, the time increment was initially set to 1.0 s, but was automatically shortened to 1.0 ms after any of the elements had deformed for their sectional width in a single step. The analysis required approximately 8 h of simulation time using a standard personal computer (CPU: 2.93 GHz Xeon).

6.3 NUMERICAL RESULTS

Fig. 6.5 shows the collapse sequences of the building when exposed to each fire pattern. In the analyses, the member joint strength ratio C_M, defined in the previous chapter, was set to 0.3 and the axial force ratio in

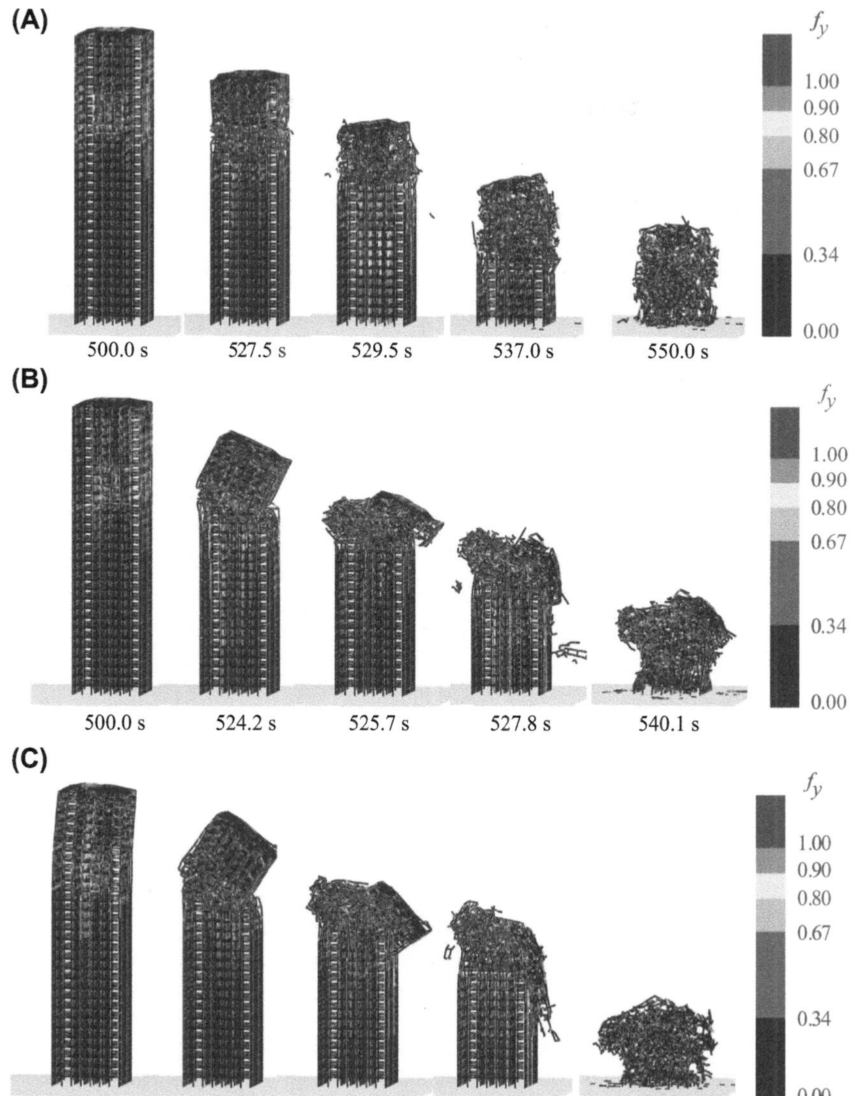

Figure 6.5 Progressive collapse sequence (axial force ratio = 0.33, $C_M = 0.3$, with an outrigger truss system). (A) Fire pattern A, (B) fire pattern B, and (C) fire pattern C.

the columns of the first floor was set to 0.33, with a regular-strength outrigger truss system placed on the roof (the structural strength was similar to that of the beams on the first floor). A total collapse was inevitable for all the cases, and the results clearly show the effect of the weak member connections compared to the fully rigid connections and the effect of the reduction in strength due to the elevated temperature. As shown in Fig. 6.5, the collision of the upper structure with the lower structure creates shock waves that propagates through the columns and moves faster than the falling motion of the upper structure. The shock waves seem to destroy the member joints as they move toward the ground level. The differences in fire patterns do not appear to affect the overall collapse behavior. However, the slant angles at the initiation of the collapse phase are different, and the time until the initiation of the collapse is shorter if the fire pattern is asymmetrical. The amount of time until the initiation of the collapse depends on the strength and the number of outrigger trusses that support the load paths near structural deficiencies. Moreover, some structural members appear to fall at the speed of a free fall, whereas the upper structure experiences a reduction in speed during the fall. This phenomenon appeared in all the cases, including the case where $C_M = 0.1$.

Fig. 6.6 shows the relation between the axial force ratio and the collapse initiation time: the duration from the beginning of the fire until collapse initiation. The amount of time was not affected by the presence of the outrigger truss system if the axial force ratio was large. However, the effect of the outrigger truss system on the collapse initiation time increased in all cases if the axial force ratio had smaller values. Therefore, the influence of the structural parameters on the collapse initiation time was investigated in two different cases: (1) axial force ratio = 0.33 (floor load: 10.0 kN/m^2), for which the difference in collapse initiation times was small, and (2) axial force ratio = 0.19 (floor load: 5.0 kN/m^2), for which the difference was larger. Fig. 6.7 shows the results for the former case. The results without plots are the cases in which the building model does not collapse during the entire calculation time. Although the strength of the outrigger trusses and

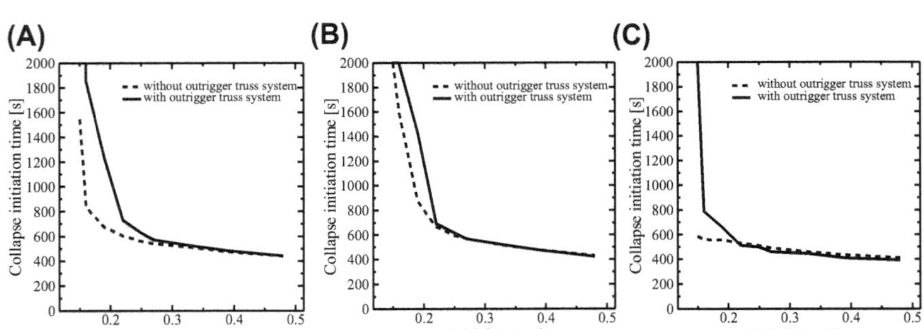

Figure 6.6 Relation between axial force ratio and collapse initiation time. (A) Fire pattern A, (B) fire pattern B, and (C) fire pattern C.

the member joints slightly increases the collapse initiation time for fire patterns A and B, it does not seem to significantly affect the collapse initiation time if the fire pattern becomes as asymmetrical as fire pattern C. In particular, all cases, except the case where $C_M = 0.1$, had a constant collapse initiation time of around 450 s when fire pattern C was applied (Fig. 6.7C). Fig. 6.8 shows the results for the latter case (2), in which the axial force ratio was smaller. The collapse initiation time seems to be largely affected by the presence of the outrigger truss system. However, the most significant parameter in this case is the member joint strength ratio, C_M, and not the member strengths of the outrigger trusses. For example, the collapse initiation time with $C_M = 0.2$ under fire pattern B varies by approximately 40 s between the models with and without the outrigger truss system, but with $C_M = 0.3$, the collapse initiation time increases for more than 520 s if the outrigger truss system is present.

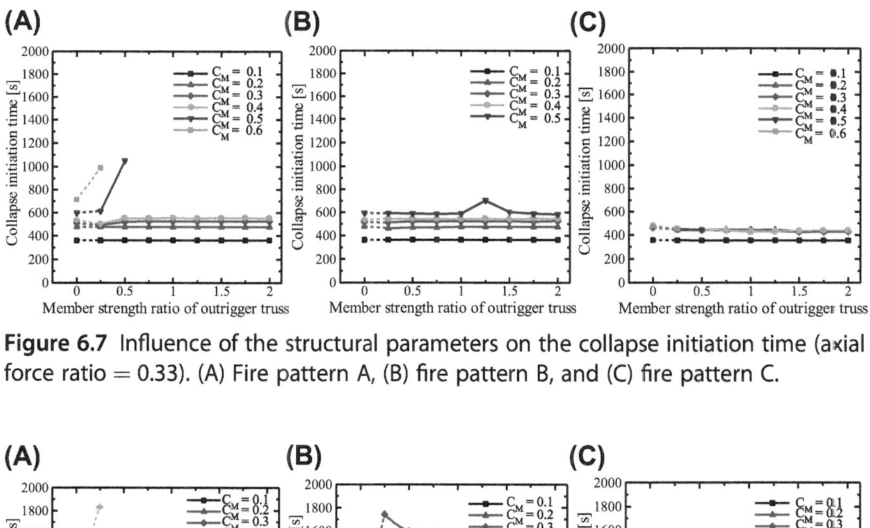

Figure 6.7 Influence of the structural parameters on the collapse initiation time (axial force ratio = 0.33). (A) Fire pattern A, (B) fire pattern B, and (C) fire pattern C.

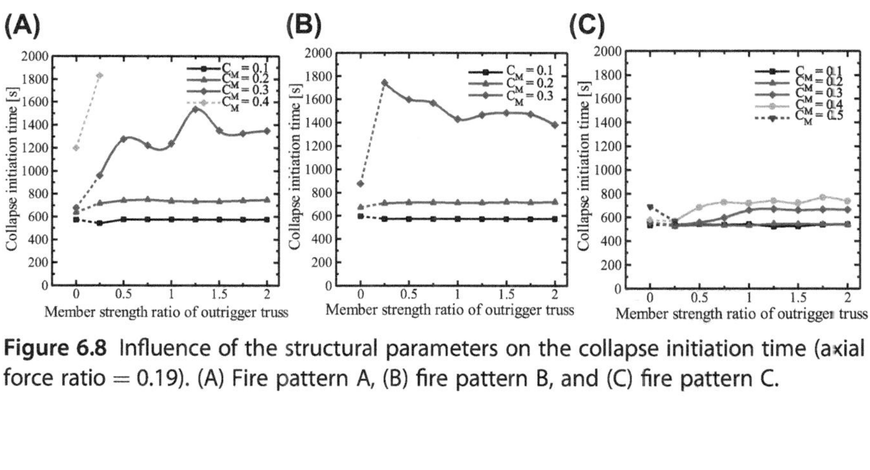

Figure 6.8 Influence of the structural parameters on the collapse initiation time (axial force ratio = 0.19). (A) Fire pattern A, (B) fire pattern B, and (C) fire pattern C.

These results indicate that the collapse initiation time can be significantly affected by the member joint strength ratio, particularly if the axial force ratio is small (floor loads are low), under the condition that the fire pattern is nearly symmetrical and the load paths to and from the outrigger truss system are sufficiently protected.

6.4 SUMMARY

The results described in this chapter have shown that the outrigger truss system delays the collapse initiation time in the models with smaller axial force ratios, greater connection strengths and symmetrical fire patterns. The results explain how WTC 1 (the north tower) in 9/11 incident, which had symmetrical structural deficiencies, was able to remain standing for a longer period of time before its total collapse compared to WTC 2, which had asymmetrical structural deficiencies. However, the collapse speed in the analyses never reached a value as high as that of the free fall observed in the WTC collapse, when the splices between column sections had at least some tensile strength remaining.

By collecting the information obtained in this chapter and Chapter 5, the following sequence regarding the total collapse of the WTC towers can be assumed. Once the partial destruction was initiated by the buckling and strength reduction of the members due to elevated temperatures, the collapse progressed like a chain reaction due to the original weakness of the member connections and the "spring-back" destruction of the splices between the sections of columns that were directly caused by the aircraft impact itself, which led, eventually, to a high-speed total collapse.

REFERENCES

Hirashima, T., Hamada, N., Ozaki, F., Ave, T., Uesugi, H., 2007. Experimental study on shear deformation behavior of high strength bolts at elevated temperature. Journal of Structural and Construction Engineering AIJ 621, 175–180 (in Japanese).

Isobe, D., Thanh, L.T.T., Sasaki, Z., 2012. Numerical simulations on the collapse behaviors of high-rise towers. International Journal of Protective Structures 3 (1), 1–19.

NIST NCSTAR 1, 2005. Federal Building and Fire Safety Investigation of the World Trade Center Disaster: Final Report on the Collapse of the World Trade Center Towers. National Institute of Standards and Technology (NIST).

CHAPTER SEVEN

Risk Estimation for Progressive Collapse of Buildings

7.1 INTRODUCTION

In case of emergency, such as intense fire or collision of some objects to buildings, there is a risk of progressive collapse; a phenomenon which occurred to the World Trade Center towers during the 9/11 terrorist attacks. The official statements released by the Federal Emergency Management Agency (FEMA) in 2002 (ASCE/FEMA, 2002), and also by the National Institute of Standards and Technology (NIST) in 2005 and 2008, concluded that the details of the failure process after the decisive initial trigger setting the upper part in motion were very complicated and their clarification would require large computer simulations. Among various papers published after the incident, the one published by Bazant and Zhou (2002) had become a milestone for other numerous works concerning progressive collapse of buildings. By defining an overload ratio—a ratio of the elastic strength of the lower structure to the total weight of the upper, falling structure—they explained that the force applied to the lower structure by the upper was overwhelmingly exceeding the design vertical load, and consequently, it must had been practically impossible to design a high-rise tower to avoid a progressive collapse that happened to the twin towers. On the other hand, they, too, admitted that numerous computer simulations should be carried out to clarify the collapse process, since the estimations were done without considering tilting of upper structure, member fracture, and other details.

In this chapter, the collapse behaviors of steel-framed buildings are simulated using the ASI-Gauss code to investigate the relation between a key element index, which indicates the contribution of a structural column to the vertical capacity of the structure, and the scale of progressive collapse. Collapse was initiated by removing specific columns from the models designed based on different axial force ratios. Some patterns of removed columns, of which the locations were restricted to a single floor, were investigated. The total potential energy values of structural members after the collapse were used to estimate the collapse scale of the buildings.

Progressive Collapse Analysis of Structures
ISBN: 978-0-12-812975-3
http://dx.doi.org/10.1016/B978-0-12-812975-3.00007-4

7.2 KEY ELEMENT INDEX

Due to variations in their span lengths and strengths, certain columns in a structure support more vertical loads than other columns and thus act as "key elements" in the structure. Several indexes have been developed to evaluate the contribution of each column to the strength of a structure, i.e., to determine the key elements, such as the redundancy index (Frangopol and Curley, 1987) and the sensitivity index (Ito et al., 2005). These indexes are effective for identifying highly sensitive columns that support the vertical load; however, they equivalently identify those columns that only cause partial collapse at the upper floors and thus they are not useful to estimate the scale of progressive collapse. Therefore, a new index called the key element index for estimating the contribution of base columns to the total collapse of the structure was proposed (Isobe, 2014).

The key element index is calculated as: First, a static pushdown analysis is performed, as shown in Fig. 7.1, by equally applying incremental vertical loads at every structural joint in a modeled structure. Several columns on the upper floors may yield in this process, but the vertical loads are nevertheless continuously applied in steps until one of the base columns yields. The total vertical load (including the floor loads) applied to a column at the step when one of the base columns yields is defined as the ultimate yield strength P_G of the structure. The ultimate yield strength of the initial, undamaged structure is denoted by ${}_0P_G$, and the ultimate yield strength of a structure with one base column eliminated is denoted by ${}_1P_G$. The key element index KI of column m, for example, can be defined as the ratio of the ultimate yield strengths of the present and the initial step. The index can thus be written as:

$$ {}_1^0KI_m = {}_0P_G / {}_1P_G \tag{7.1} $$

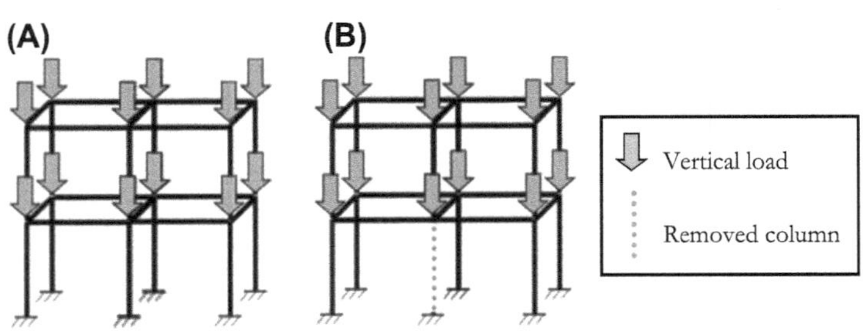

Figure 7.1 Pushdown procedure for calculating key element index. (A) Initial model and (B) model with one column removed.

The key element index at the nth step is then written as:

$$ {}_{n}^{0}KI_{m} = {}_{0}P_{G} / {}_{n}P_{G} \tag{7.2} $$

This index indicates the contribution of the present column (or columns) to the strength of the initial, undamaged structure and can be used for cases in which the columns are removed simultaneously. However, the sensitivity of columns may change momentarily when the columns are removed sequentially within a certain interval (such as blast demolition cases shown in the next chapter). The updated index, as shown below, is used for the cases with a sequential removal:

$$ {}_{n}^{n-1}KI_{m} = {}_{n-1}P_{G} / {}_{n}P_{G} \tag{7.3} $$

In this case, the index can be defined as the ratio of the ultimate yield strengths of the present and the previous steps. An example of the distribution of key element index values in a building model can be found later in Section 7.5.

7.3 NUMERICAL MODELS AND CONDITIONS

A 10-story, 3-span, steel-framed building model, as shown in Fig. 7.2, was constructed for the progressive collapse analysis. Each span length of the model was 7 m, and the height between each floor was 4 m. The columns were made of SM490 steel with box-type sections and the beams were made of SM400 steel with H-type sections. The material properties are shown in Table 7.1. The columns were subdivided with two linear Timoshenko beam elements per member, and the beams and floors were subdivided

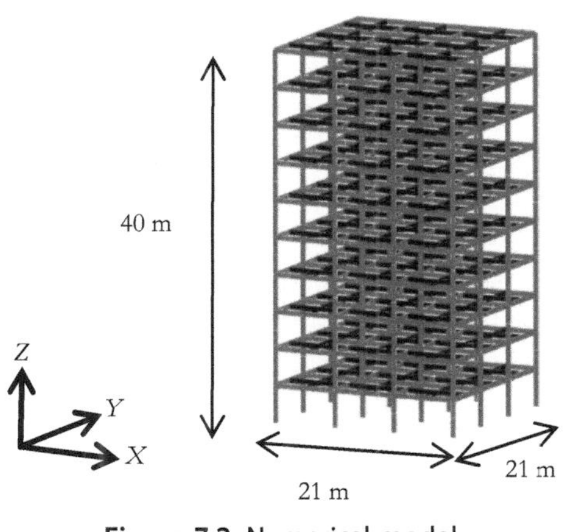

Figure 7.2 Numerical model.

Table 7.1 Material Properties of Steel Members

	E	σ_y	ν	ρ
SM400	206	245	0.3	7.9×10^{-6}
SM490	206	325	0.3	7.9×10^{-6}

E, Young's modulus (GPa); σ_y, yield stress (MPa); ν, Poisson's ratio; ρ, density (kg/mm^3).

with four elements. The floor elements were all modeled with elastic elements that do not yield. Several models with different strengths were prepared, as shown in Table 7.2, to see the effects of structural strengths on the collapse scale of buildings. An 800 kgf/m^2 of floor load was taken into account, and the sectional sizes of columns and beams were decided according to the base shear coefficients of buildings when designing the models. Each model is notated as Models A, B, C, D, and E, from the strongest to the weakest.

The collapse modes of buildings were expected to change according to their number of removed columns and locations. Thus, the patterns of removed columns were selected as shown in Fig. 7.3, and the removal of columns was restricted to a single floor for simplicity. An analysis was carried out for 10.0 s of simulation time with a time increment of 1.0 ms. The removal of columns was executed at $t = 1.0$ s in the simulations.

A decrease ratio of potential energy, as defined as the following equation, was used to evaluate the scale of progressive collapse:

$$\text{Decrease ratio of potential energy} = \frac{U_0 - U_f}{U_0} \tag{7.4}$$

where U_0 and U_f are the potential energies of numerical models at the initial and final states, respectively. The potential energy U is defined as the sum of potential energies of all elements composing the model and is expressed as:

$$U = \sum_{i=1}^{i_M} mgh_i \tag{7.5}$$

Table 7.2 Structural Strengths of Each Model

	Maximum Axial Force Ratio n	**Base Shear Coefficient C_b**
Model A	0.124	0.200
Model B	0.200	0.095
Model C	0.300	0.048
Model D	0.400	0.027
Model E	0.500	0.016

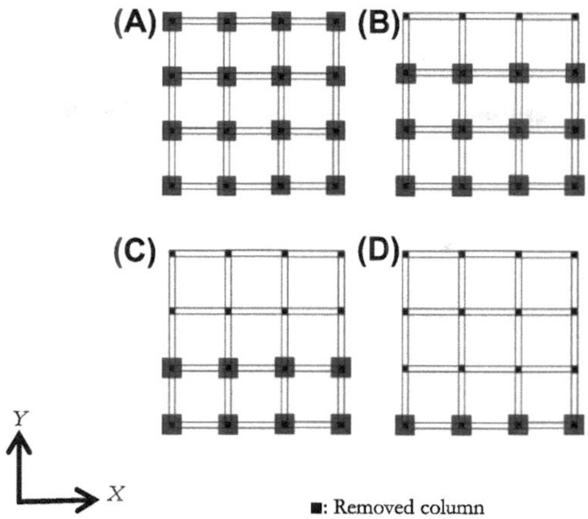

Figure 7.3 Removed-column patterns. (A) Pattern 1, (B) Pattern 2, (C) Pattern 3, and (D) Pattern 4.

where i, i_M, are the element numbers, excluding the fractured ones, and m, g, and h_i are mass of the element, gravitational acceleration, and height from the ground to the center of the element number i, respectively. The scale of progressive collapse is overestimated, thus it can be underestimated on a safety point of view, by applying Eq. (7.5) only to non-fractured elements. The value calculated by Eq. (7.4) indicates a larger scale of progressive collapse if the value is nearer to 1.0.

7.4 PROGRESSIVE COLLAPSE BEHAVIORS OF A STEEL-FRAMED BUILDING

As mentioned in the previous section, the patterns shown in Fig. 7.3 were used to investigate the collapse behaviors of the models. Total of 200 cases of simulations were carried out by varying the floor of removed columns from the 1st floor to 10th floor.

The collapse sequences of the models are shown in Fig. 7.4. First, Fig. 7.4A and B show the results for Models C and E when the columns are removed with Pattern 1 at the fourth floor. Shock waves propagating throughout the building are observed after the collision between the upper and lower structures occurs at $t = 1.9$ s. Then, Model C with relatively strong structural strength withstands and stops the collapse from progressing. On the other hand, Model E with relatively weak structural strength cannot withstand the impact, and finally, advances to a total collapse. Asymmetricity appears when columns are removed with Pattern 2 at the 7th floor for the

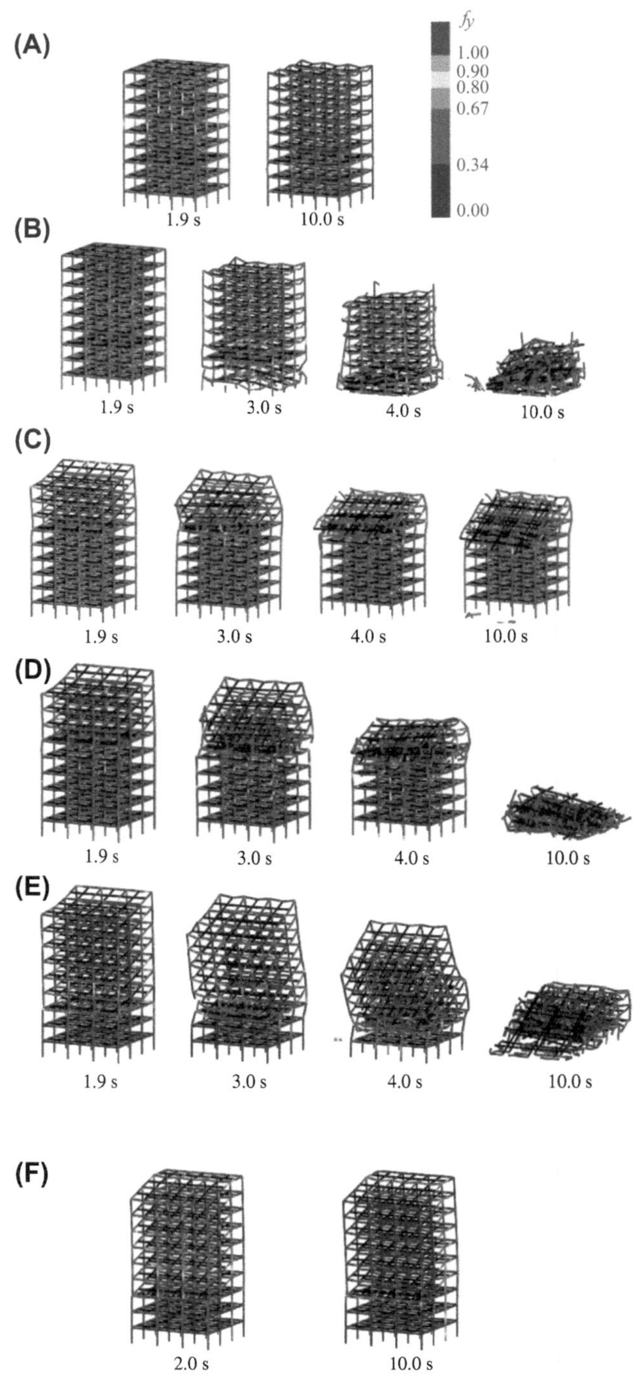

Figure 7.4 Collapse sequences under various conditions. (A) Model C, Pattern 1 at fourth floor, (B) Model E, Pattern 1 at fourth floor, (C) Model C, Pattern 2 at seventh floor, (D) Model E, Pattern 2 at seventh floor, (E) Model C, Pattern 2 at fourth floor, and (F) Model C, Pattern 3 at fourth floor.

same models (Fig. 7.4C and D). Although the upper structure in Model C crashes after the impact, it does not advance to a progressive collapse. On the contrary, the tilting of the upper structure gives severe damages to the lower structure on impact, in Model E, and triggers a total collapse. Fig. 7.4E and F show the cases when columns are removed with Patterns 2 and 3 at the same fourth floor for Model C. Here, it is confirmed that the collapse stops from progressing when the number of removed columns decreases from Pattern 2 to Pattern 3. However, the risk of progression does not necessarily increase in accordance with the number of removed columns; the collapse stopped from progressing in Fig. 7.4A when all 16 columns are removed. There might be a relevance to the asymmetricity of the removed-column patterns. Furthermore, it can be clearly confirmed by comparing Fig. 7.4C and E that the risk of progression becomes higher if the removed columns are located in lower floors.

The numerical results of 200 cases indicate that the collapse modes tend to vary depending on the structural strengths of the models, the floor of the removed columns and the removed-column patterns. Fig. 7.5 shows the

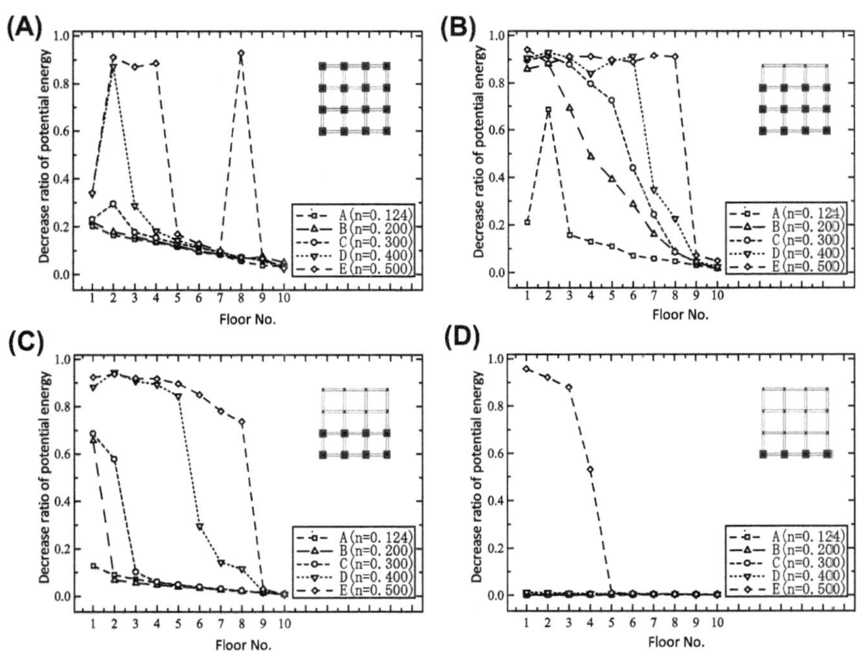

Figure 7.5 Relation between the Floor No. where columns are removed and the decrease ratio of potential energy for each pattern. (A) Pattern 1, (B) Pattern 2, (C) Pattern 3, and (D) Pattern 4.

relation between the Floor Nos. where columns are removed and the decrease ratio of potential energy for each pattern. There is a clear dependency of the decrease ratio to the location of the floor where columns are removed in Patterns 2, 3, and 4 (Fig. 7.5B–D); the lower the floor, the higher the ratio. Thus, it indicates a high risk of large-scale collapse when failure of columns occurs at lower floors. This is due to the reason that the impact force from upper to lower structure increases due to larger kinetic energy of the upper structure, and also to the tendency of buildings becoming more unstable if the columns at lower floors are removed. The figure also shows that the decrease ratio increases as the strength of the model becomes weaker from Model A to Model E, and as the number of removed columns becomes larger. However, there is a slight difference in this tendency in case of Pattern 1, as shown in Fig. 7.5A. Although the number of removed columns is larger than in Pattern 2, the removed-column pattern is fully symmetric and stable cases that do not lead to a large-scale collapse tend to appear more often. The impact between the upper and lower structures occurs in a wider area with less tilting in this case, and impact force is distributed in wider range compared to the other patterns, which should produce the stability.

7.5 RISK ESTIMATION FOR PROGRESSIVE COLLAPSE USING KEY ELEMENT INDEX

Next, by fixing the number of removed columns to 12, the removed-column patterns as shown in Fig. 7.6 were selected to investigate the relation

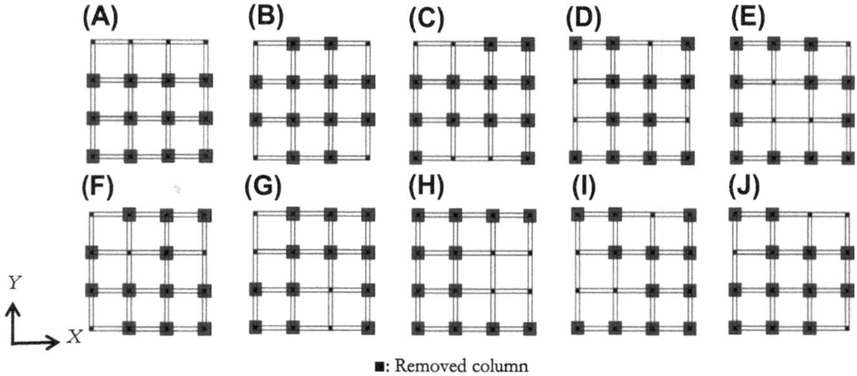

Figure 7.6 Removed-column patterns for detailed investigation. (A) Pattern 2-1, (B) Pattern 2-2, (C) Pattern 2-3, (D) Pattern 2-4, (E) Pattern 2-5, (F) Pattern 2-6, (G) Pattern 2-7, (H) Pattern 2-8, (I) Pattern 2-9, and (J) Pattern 2-10.

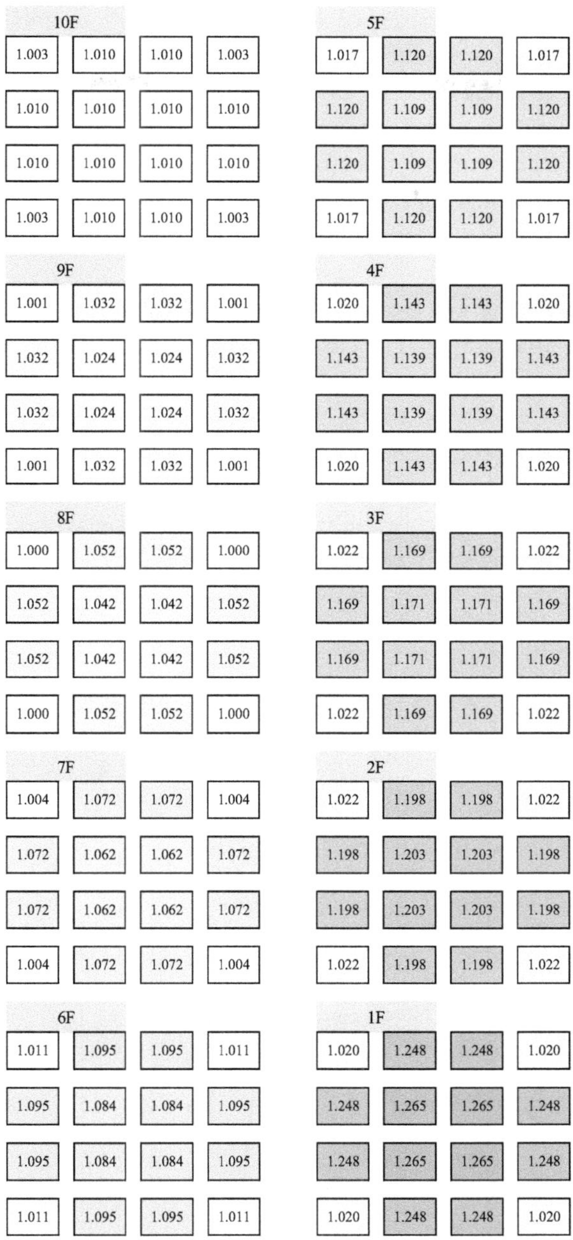

Figure 7.7 Key element index value of each column for Model A.

between the integrated values of *KI* and the risk of progressive collapse. Pattern 2-1 in the figure is the same pattern as Pattern 2 used in the previous section and it should be the most asymmetric pattern in Fig. 7.6. Pattern 2-2, on the other hand, should be the most symmetric pattern of all. Other patterns were selected randomly. Asymmetricity of the patterns were evaluated by calculating the moving distance of the center of gravity in horizontal plane before and after the removal of columns, and the effect against collapse scale was also investigated.

Fig. 7.7 shows the key element index values calculated for each column in Model A, as an example. The columns with larger *KI* values are illustrated with deeper gray in the figure. The columns in the corner have smaller values and the ones at the lower floors have larger values. Other models also had the same tendencies. The integrated values of *KI* are calculated by summing up all the *KI* values of removed columns.

Fig. 7.8 shows the relation between the Floor Nos. where columns are removed and the decrease ratio of potential energy, for Models A, C, and E.

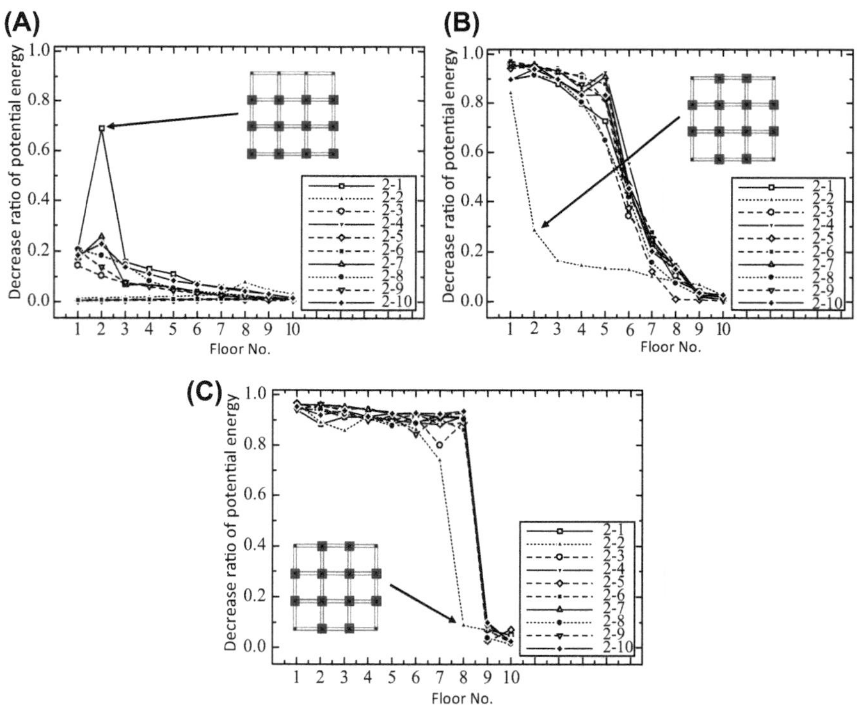

Figure 7.8 Relation between the Floor No. where columns are removed and the decrease ratio of potential energy for each model. (A) Model A, (B) Model C, and (C) Model E.

The tendencies of the plots are same as in the previous section; the lower the strength of the model and the lower the floor where columns are removed, the higher the ratio or the risk of large-scale collapse. If the results are surveyed for each model, a variation in the ratio depending on the removed-column patterns can be observed. Particularly, a large-scale collapse occurred only in case of Pattern 2-1, the most asymmetric pattern of all, in Model A, the strongest model of all. On the contrary, the collapse scale became smaller in case of Pattern 2-2, the most symmetric pattern of all, in Models C and E, the weaker models.

Fig. 7.9 shows the relation between the integrated values of *KI* and the decrease ratio of potential energy, for Models A, C, and E. It is evident that as the integrated values of *KI* becomes larger, the ratio, thus the collapse scale, becomes larger. Moreover, the critical integrated values of *KI* where large-scale collapse begins, becomes smaller as the strength of the model becomes weaker from Model A to Model E. This means that there are some critical integrated value of *KI* to start a large-scale collapse, which depend on

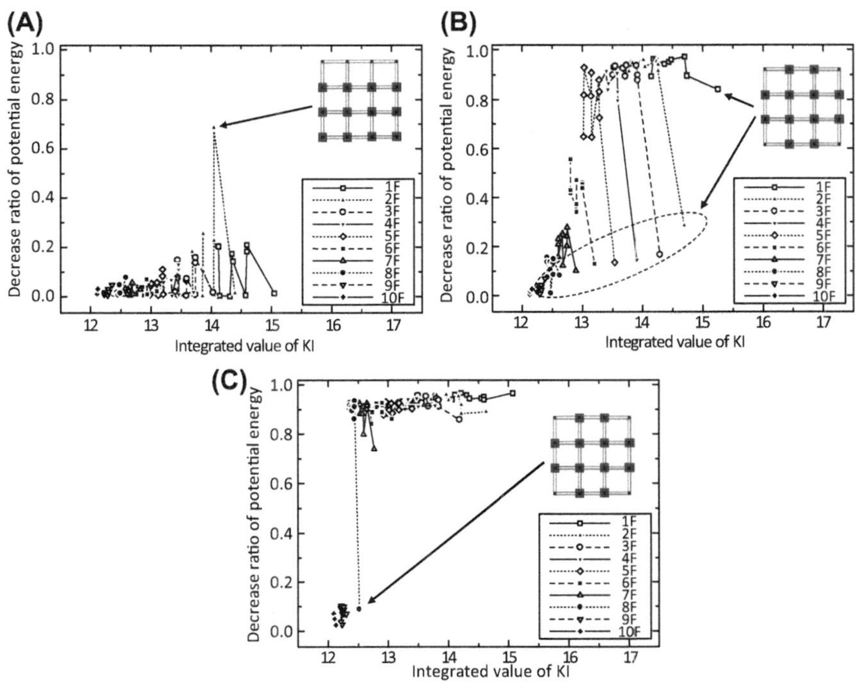

Figure 7.9 Relation between the integrated values of *KI* and the decrease ratio of potential energy for each model. (A) Model A, (B) Model C, and (C) Model E.

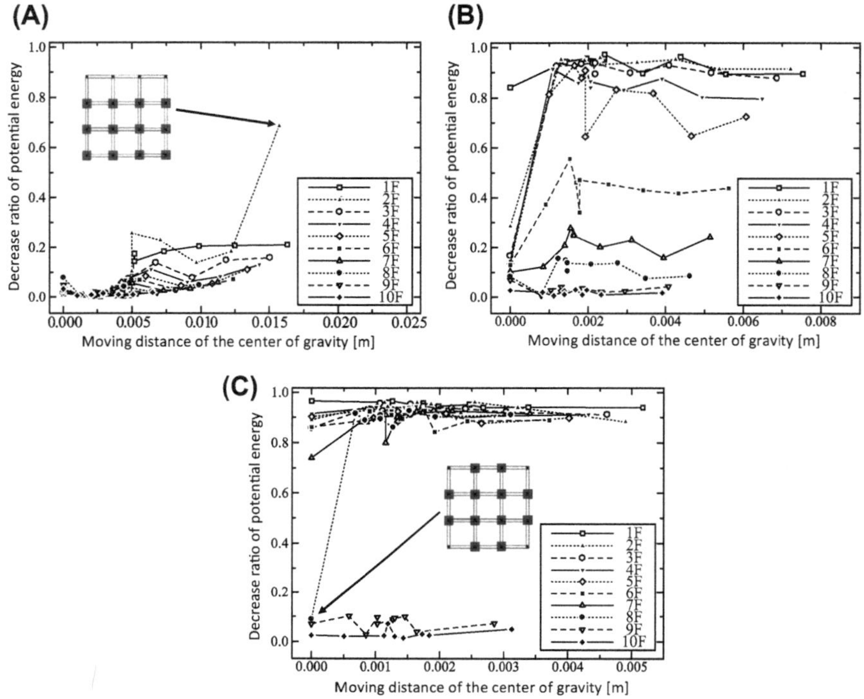

Figure 7.10 Relation between the moving distance of the center of gravity and the decrease ratio of potential energy for each model. (A) Model A, (B) Model C, and (C) Model E.

the strengths of models. The conditions of Pattern 2-1 on second floor of Model A, and Pattern 2-2 on some cases of Models C and E, also show some irregular tendencies as observed in Fig. 7.8.

To see the effect of the asymmetricity of the building before and after the removal of columns, the relation between the moving distance of the center of gravity in horizontal plane and the decrease ratio of potential energy is investigated, as shown in Fig. 7.10. The mass of the removed columns are very small compared to the rest of the building, and the moving distance due to the removal is trivial. However, it can be observed that the collapse scale becomes larger in accordance with the moving distance in horizontal plane. It is also to be noted that the moving distances of the center of gravity for Patterns 2-1 and 2-2 are located at extreme positions in the figures, which explains the results shown in Figs. 7.8 and 7.9. The collapse scale of buildings are largely affected by the asymmetricity, or the symmetricity, of the structural failure.

7.6 SUMMARY

By evaluating the numerical results using the key element index, it was found that the larger the integrated value of key element index, the higher the risk of progressive collapse; however, some peculiar tendencies were observed in the cases of removed columns with extremely symmetrical or asymmetrical locations. The critical integrated value of the key element index needed to cause a large-scale progressive collapse tends to depend on the strength of the building, and the value cannot be generally determined. This is due to the fact that the key element index is a parameter which does not relate to the strength of the building itself, and it cannot be compared relatively between buildings with different strengths. Therefore, the key element index may be used to predict and compare the risks of progressive collapse, only in the same building, but even when the locations of removed columns are variously assumed.

REFERENCES

ASCE/FEMA, 2002. World Trade Center Building Performance Study: Data Collection, Preliminary Observation and Recommendations. ASCE/FEMA.

Bazant, Z.P., Zhou, Y., January 2002. Why did the world trade center collapse? —simple analysis. Journal of Engineering Mechanics 2—6.

Frangopol, D.M., Curley, J.P., 1987. Effects of damage and redundancy on structural reliability. Journal of Structural Engineering, ASCE 113 (7), 1533—1549.

Isobe, D., 2014. An analysis code and a planning tool based on a key element index for controlled explosive demolition. International Journal of High-Rise Buildings 3 (4), 243—254.

Ito, T., Ohi, K., Li, Z., 2005. A sensitivity analysis related to redundancy of framed structures subjected to vertical loads. Journal of Structural and Construction Engineering, AIJ 593, 145—151 (in Japanese).

NIST NCSTAR 1, 2005. Federal Building and Fire Safety Investigation of the World Trade Center Disaster: Final Report on the Collapse of the World Trade Center Towers. National Institute of Standards and Technology (NIST).

NIST NCSTAR 1A, 2008. Final Report on the Collapse of World Trade Center Building 7. National Institute of Standards and Technology (NIST).

CHAPTER EIGHT

Blast Demolition Analysis of Buildings

8.1 INTRODUCTION

The technology used in the demolition of old or uninhabitable buildings has always been of major interest in civil engineering and remains a challenge in engineering practice. Demolition techniques using controlled explosives are often used to increase work efficiency, because in general, conventional demolition techniques that use a hydraulic concrete crusher, a concrete cutter or a nonexplosive demolition agent are lengthy and costly. However, the former techniques pose a high risk of damages to neighboring buildings especially in urban areas (Williams, 1990), and require high levels of knowledge and experience (Yarimer, 1989; Kinoshita et al., 1989). Although there are several explosive demolition companies currently operated in the United States and Europe, explosive demolition techniques are not popularly used in Japan due to the restriction on using large amount of explosives and the strong, anti-seismic design strengths of the buildings. In these circumstances, assumptions using numerical simulations can be helpful in ensuring the success of the technique, even for the countries with high restrictions.

In this chapter, a numerical code developed using the ASI-Gauss technique is validated by comparing the numerical results with a simulated demolition experiment. Then, a demolition planning tool based on the key element index (*KI*) described in the previous chapter was developed. Two ways of selecting specific columns to demolish the entire structure are demonstrated: selecting the columns from the largest index value and from the smallest index value. The demolition results were confirmed numerically by carrying out collapse analyses using the ASI-Gauss code, and the tendencies of the demolition modes to follow the *KI* values were estimated (Isobe, 2014). Furthermore, some other numerical examples of blast demolition analyses are shown.

8.2 VALIDATION OF THE METHODS BY EXPERIMENTS

The numerical code was validated by comparing the results obtained from an experiment performed in the laboratory using an electromagnetic

Progressive Collapse Analysis of Structures
ISBN: 978-0-12-812975-3
http://dx.doi.org/10.1016/B978-0-12-812975-3.00008-6

system to experimentally simulate the explosive demolition of a structure (Isobe et al., 2006). The outline of the experimental system is shown in Fig. 8.1. Two electromagnetic devices (KANETEC, KE-4RA, max. attach force = 110 N, for beams; KE-4B, max. attach force = 400 N, for columns) are mounted on both sides of an aluminum pipe, and the members are joined with steel connectors held by the magnetic field generated by the electromagnetic devices. Power is supplied to these devices through a 4-pin connector jack and a telephone cord. The blast intervals and switching of the magnetic field are controlled by the blast interval controller and power switches, which are connected to a PC and a power supply. The magnetic fields of the devices at each blast point are sequentially released by the PC to simulate member fracture caused by an explosion. The system can simulate the failure of beams by disconnecting the magnetic force between the beams and the columns. However, the column failure cannot be initiated in the electromagnetic system as in the real conditions. The failures of the columns are simulated, first by switching off the electromagnetic field of the column member joints and then, by composing a cantilever system with the upper beam. The purpose of this system is to pile up the demolition test results, easily and safely, to validate the numerical code.

The properties and conditions used for the experiment and the analysis are as follows: each span length is 34 cm for the beams and 28 cm for the columns, with 3 cm × 3 cm box-type beams and 4 cm × 4 cm box-type

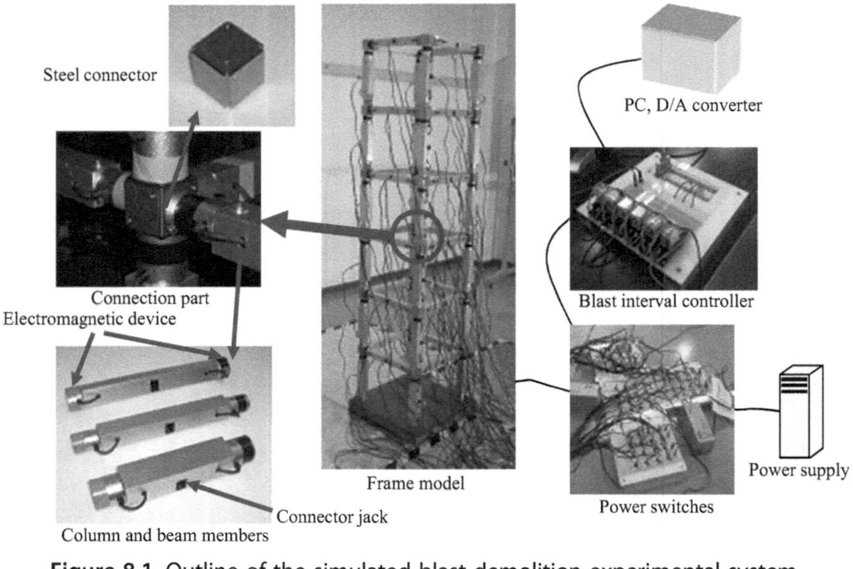

Figure 8.1 Outline of the simulated blast demolition experimental system.

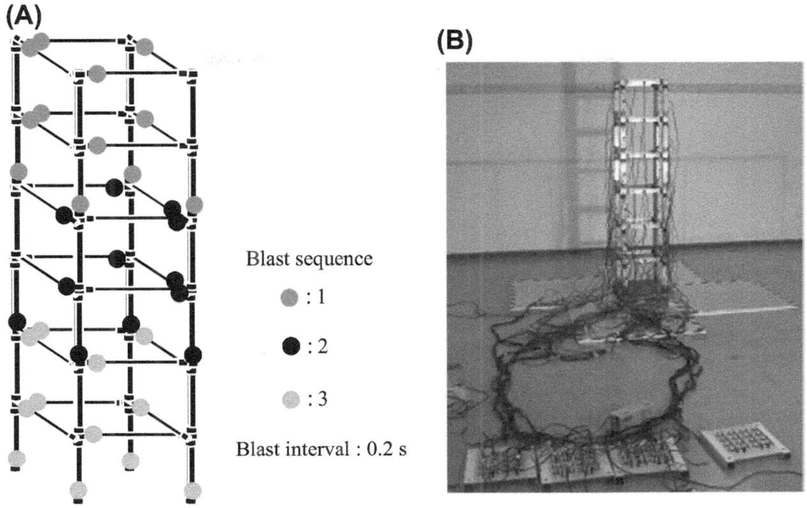

Figure 8.2 Blast conditions for a six-story frame model. (A) Blast conditions and (B) experimental setup.

columns, each with a Young's modulus of 70 GPa and a Poisson ratio of 0.345 (aluminum); the critical values of the bending strains are 2.4×10^{-4} rad/mm (beams) and 3.1×10^{-4} rad/mm (columns); and the tension strains are 2.7×10^{-4} (beams) and 3.0×10^{-4} (columns). A time increment of 0.1 ms was used for the analysis. Structural damping was not considered; however, numerical damping was considered in the time integration using Newmark's β method.

A six-story frame model was constructed using the described devices. The blast conditions for the model are shown in Fig. 8.2; the blast points and the blast sequence were carefully selected to enable the floor sections to fold in sequentially starting at the upper floors so as to avoid the dispersal of fragments over a wide area. Fig. 8.3 shows a comparison of the experimental demolition modes and the numerical results. The expected demolition mode can be observed in both cases, and the results are in very good agreement.

8.3 BLAST DEMOLITION PLANNING TOOL USING THE KEY ELEMENT INDEX

In this section, a blast demolition planning tool using the *KI* is first described. Then, several numerical results on a 15-story, 3-span framed structure using the demolition plans obtained by the planning tool are shown to demonstrate its utility.

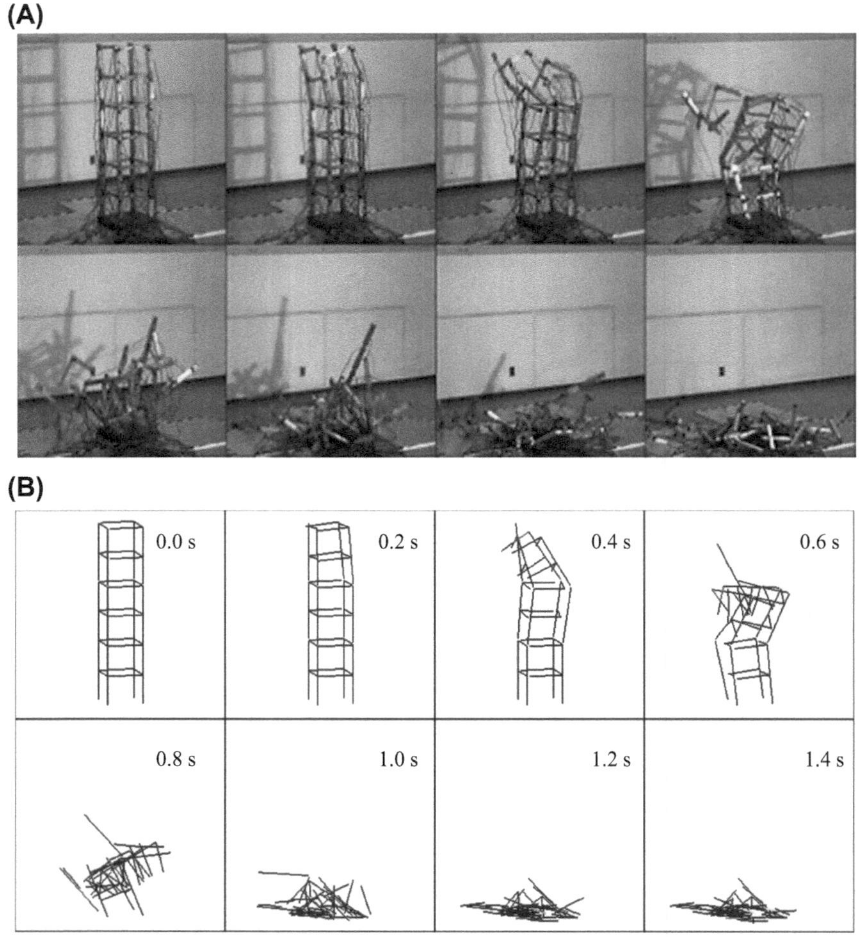

Figure 8.3 Comparison of demolition modes. (A) Experimental result and (B) numerical result.

8.3.1 Key Element Index Values for a Numerical Model

Following the procedure shown in the previous chapter, the *KI* values for a 15-story, 3-span model, as shown in Fig. 8.4, were calculated. Each span length of the model was 6 m, and the height between each floor was 3.6 m. The material used for the columns and beams was SN490B steel. A box-type section of 430 mm × 430 mm × 13 mm × 13 mm on the first floor columns and H-type section of 331 mm × 825.7 mm × 18.4 mm × 13.2 mm for the beams were selected, with the sectional sizes becoming gradually thinner in the higher stories. The floor load, in this case, was set

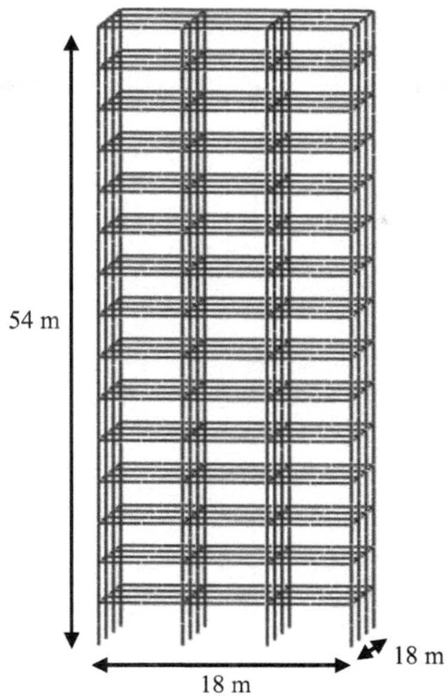

Figure 8.4 A 15-story, 3-span model.

to 400 kgf/m^2. The updated index values and the selection of columns from the largest index value at each step are shown in Fig. 8.5. For example, the *KI* value for column No. 16 (in the upper right corner) at the first step can be calculated as:

$$ {}_{1}^{0}KI_{16} = {}_{0}P_G / {}_{1}P_G = 1.693 \tag{8.1}$$

and at the second step as:

$$ {}_{2}^{1}KI_{16} = {}_{1}P_G / {}_{2}P_G = 1.012 \tag{8.2}$$

In this case, the columns at the left and lower sides were selected prior to other columns, if the index values calculated for multiple columns were the same, to direct the whole collapsing building toward the lower left side. As shown in the figure, the selected columns tend to bias to the intended demolition direction if the columns were selected from the largest index value. On the other hand, the distribution of the selected columns becomes different in pattern, as shown in Fig. 8.6, when the columns were selected from the smallest index value. The selected columns tend to distribute the load, consequently stabilizing the structure and preventing instantaneous collapse.

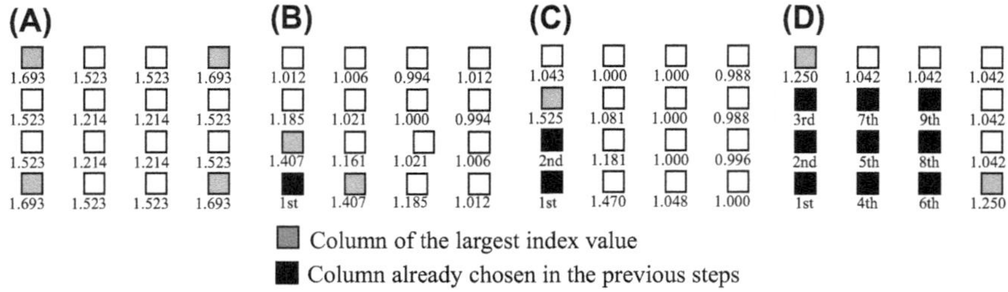

Figure 8.5 Updated *KI* values and the selection of columns from among those with the largest values (column at the left and lower side is selected prior to the other columns if they show same *KI* values). (A) 1st step, (B) 2nd step, (C) 3rd step, (D) 10th step.

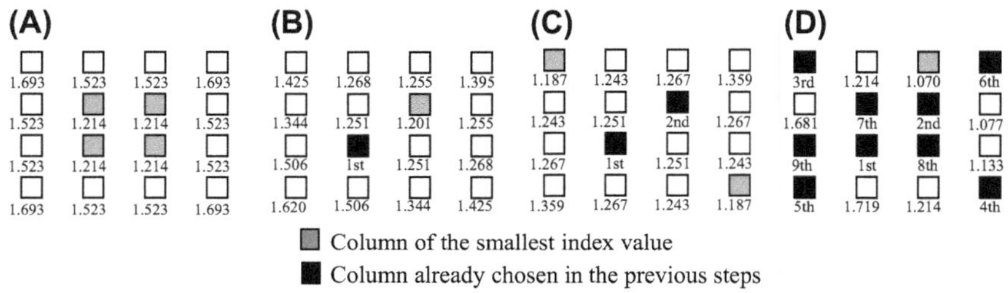

Figure 8.6 Updated *KI* values and the selection of columns from among those with the smallest values (column at the left and lower side is selected prior to the other columns if they show same *KI* values). (A) 1st step, (B) 2nd step, (C) 3rd step, (D) 10th step.

8.3.2 Blast Demolition Planning Using the Integrated Values of the Key Element Index

Several blast demolition analyses are performed using the selected columns, as described in the previous section, as the ones to be removed. The *KI* values of the selected columns were summed and the collapse modes for each case were plotted. Each procedure was performed on models with different floor loads. Fig. 8.7 shows the distribution of the integrated *KI* values and the collapse modes of the structure for a simultaneous blast, where nonupdated index values are used. TC, PC, and NC in the figures indicate total collapse, partial collapse, and no collapse, respectively. The integrated value indicated by the black circle in Fig. 8.7A, for example, is calculated as follows:

$$
\begin{aligned}
&{}^0_1KI_1 + {}^0_2KI_5 + {}^0_3KI_9 + {}^0_4KI_2 + {}^0_5KI_6 + {}^0_6KI_3 \\
&\quad = 1.693 + 2.382 + 3.633 + 4.284 + 5.979 + 7.863 = 25.834 \qquad (8.3)
\end{aligned}
$$

where the lower right subscript in the *KI* indicates the column number, the upper left subscript indicates the step number under consideration in the

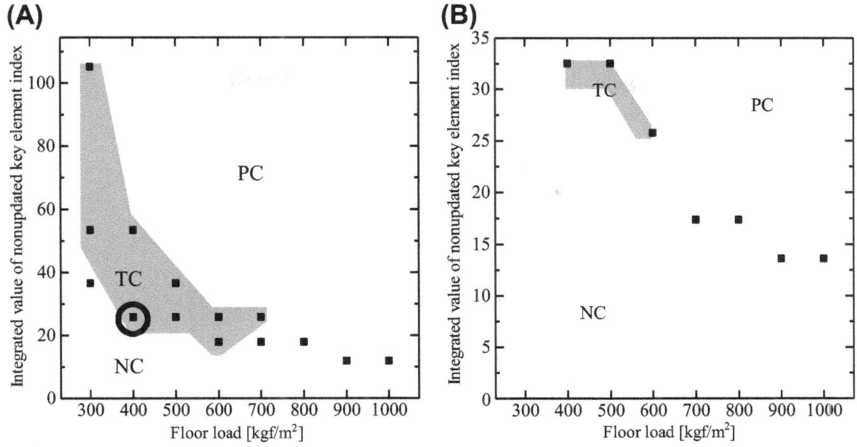

Figure 8.7 Integrated *KI* values and collapse behavior of the model (simultaneous blast, nonupdated index value). (A) Selecting columns from the largest index value and (B) from the smallest index value.

calculation of the ultimate yield strength ratio, and the lower left subscript indicates the present step number. As shown in the figures, there are certain requirements for the integrated *KI* values to ensure that the structure is partially or totally demolished. The region of TC is large when the columns with the largest index values are successively selected (see Fig. 8.7A), whereas the region is very small when the columns with the smallest index values are selected (see Fig. 8.7B). To accomplish TC, neither the integrated index values nor the floor load should be too large because larger values tend to make the structure relatively stable in the collapse process.

Fig. 8.8 shows the distribution of the integrated *KI* values and the collapse modes of the structure for sequential blast, where updated index values are used. The integrated value indicated by the black circle in Fig. 8.8A, for example, is calculated as follows:

$$ {}^{0}_{1}KI_1 + {}^{1}_{2}KI_5 + {}^{2}_{3}KI_9 + {}^{3}_{4}KI_2 + {}^{4}_{5}KI_6 + {}^{5}_{6}KI_3 + {}^{6}_{7}KI_{10} $$
$$ = 1.693 + 1.407 + 1.525 + 1.179 + 1.396 + 1.315 + 1.377 = 9.892 \tag{8.4} $$

The same tendencies as observed in Fig. 8.7 can be seen here in the distribution of the index values and the collapse modes. However, the region of TC becomes much smaller than in simultaneous blasting because the structure tends to become relatively stable during each blast interval.

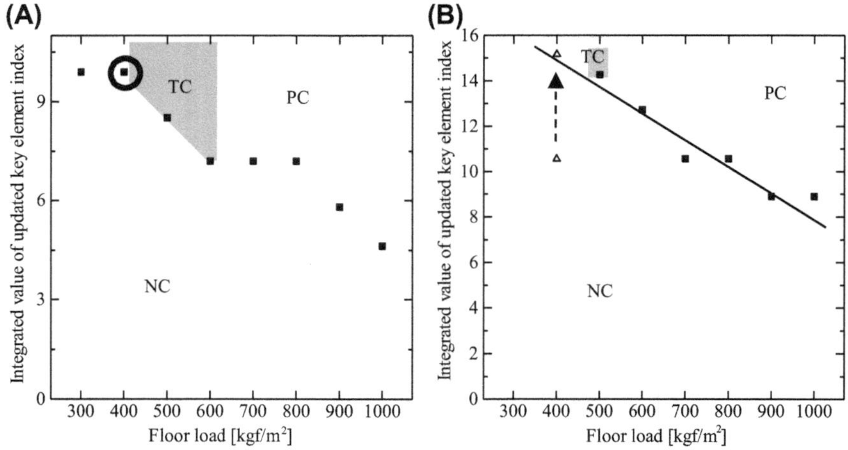

Figure 8.8 Integrated *KI* values and collapse behavior of the model (sequential blast, updated index value). (A) Selecting columns from the largest index value and (B) from the smallest index value.

It is evident from Figs. 8.7 and 8.8 that there are several specific areas of integrated index values that allow for successful demolition, especially when the columns are selected from the largest key element index values. However, the sequential blast of columns from the largest index values causes the whole structure to gradually collapse in sequence, which may, in practical terms, cut the explosive trigger wires and/or make the whole demolition procedure quite unstable. Therefore, the selection of columns with the largest index values may only be used in demolition by simultaneous blasting. In contrast, the selection of columns with the smallest index values may be used to keep the structure stable until the last moment in a demolition by sequential blasting. The structure may be completely demolished by a final blast on a column (or columns) with a large index value, essentially traversing the solid line in Fig. 8.8B in one step.

8.3.3 Blast Demolition Analysis of a Framed Structure Using the Obtained Plan

Two analyses using the plans of blast demolition obtained in the previous section were performed on the model shown in Fig. 8.4. Each analysis took 50–90 min on a personal computer (CPU: 2.0 GHz Xeon, 8 GB RAM). Fig. 8.9 shows the demolition mode for the simultaneous blast case, in which the eliminated columns were selected from among those with the largest nonupdated index values. Six columns altogether were selected, in this case, and the integrated *KI* value was 25.834, as calculated

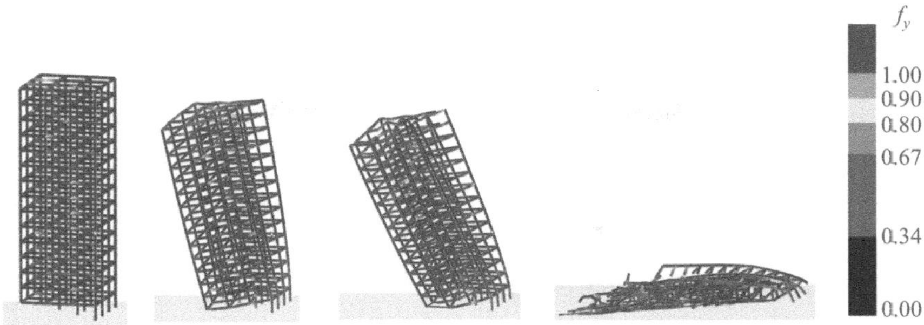

Figure 8.9 Demolition mode for the simultaneous blast case (six columns selected from the largest nonupdated index value).

in Eq. (8.3). Fig. 8.10 shows the demolition mode for the sequential blast case, in which the eliminated columns were selected from among those with the smallest updated index values. To achieve a TC, the columns with the large index values at the final step were selected for the final blast. In this case, nine columns from among those with the smallest index values (with the integrated value of 10.552) were sequentially eliminated over an interval of 1 s, and then, three columns with the large index values (with the integrated value of 4.614) were simultaneously eliminated at the final blast. The integrated index value of the initial nine columns does not traverse the solid line in Fig. 8.8B; however, the total value including the final three columns, 15.166, slightly traverses the line. Although the timings of the collapse initiation and the collapse modes slightly differ in each case, both display a successful demolition pattern, with the structure totally demolished to ground level and toward the intended direction.

Figure 8.10 Demolition mode for the sequential blast case (nine columns selected from the smallest updated index values and three columns from the large ones selected for the final blast).

8.4 OTHER NUMERICAL EXAMPLES OF BLAST DEMOLITION ANALYSIS

In this section, several other examples of blast demolition analysis are discussed.

The first analysis was carried out on a high-rise hotel, as shown in Fig. 8.11, which was structurally designed based on Japanese codes. The building was planned to be demolished in its own wake, by carefully selecting the blast time interval and the columns of specific floors to be removed. Half of the columns on either side of the floors were removed sequentially from the upper floor with a time interval of 0.5 s; 27F ($t = 0.0$ s), 19F ($t = 0.5$ s), 11F ($t = 1.0$ s), 4F ($t = 1.5$ s). As can be seen from the figure, it is successfully demolished without damaging the concourse area besides

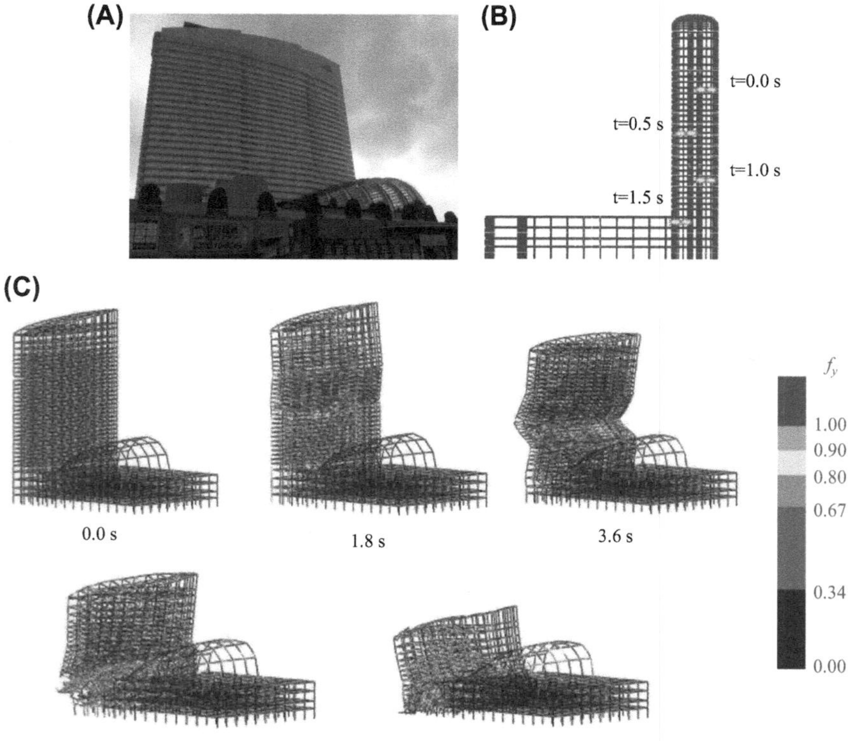

Figure 8.11 Blast demolition analysis of a high-rise hotel [columns removed sequentially from the upper floor with time interval of 0.5 s; 27F ($t = 0.0$ s), 19F ($t = 0.5$ s), 11F ($t = 1.0$ s), 4F ($t = 1.5$ s)]. (A) Analyzed hotel, (B) blast plan, and (C) demolition sequence.

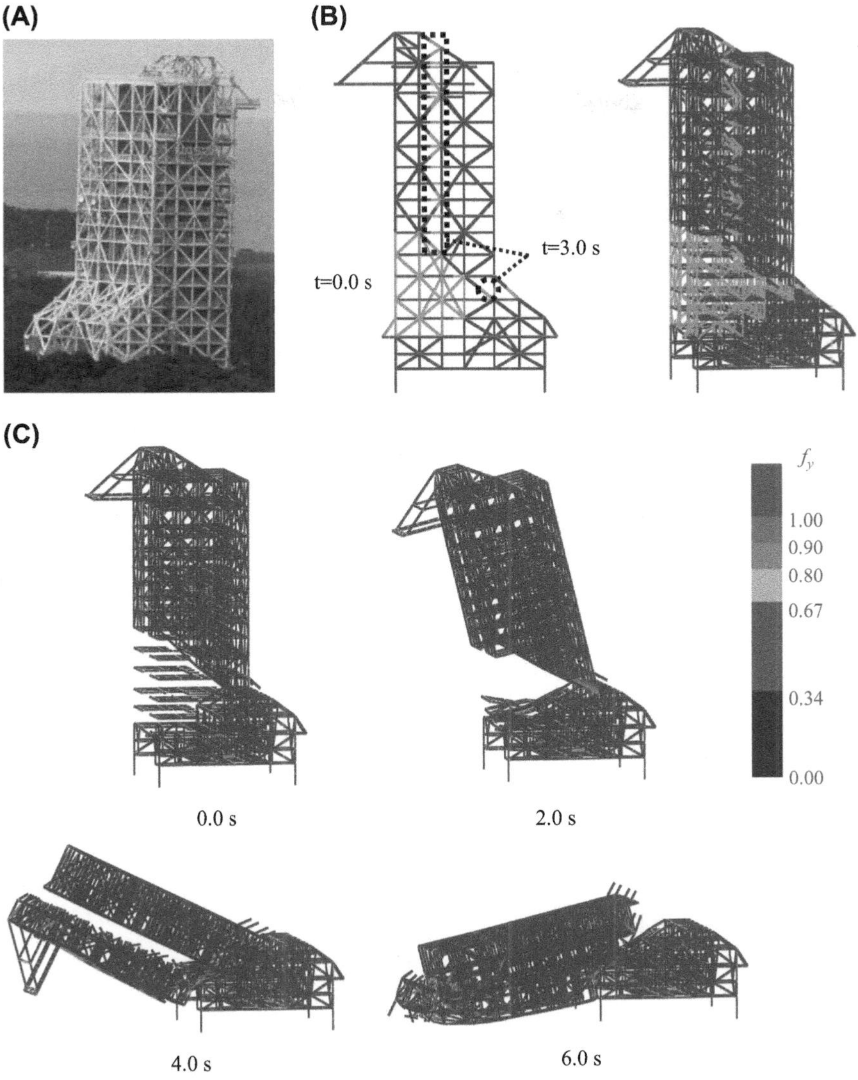

Figure 8.12 Blast demolition analysis of a rocket launch tower (sequential blasts with a time interval of 3.0 s aiming to split the upper structure). (A) Rocket launch tower (JAXA), (B) blast plan, and (C) demolition sequence.

the building. However, the high pile of remains in its wake can pose a serious problem when clearing up the area.

Second analysis was carried out on an old rocket launch tower, a very strong, steel structure with a height of over 50 m, as shown in Fig. 8.12. The demolition plan had to be carefully made, in this case too, because the post-demolition job will become more dangerous if the structure with a considerable height remained unstable. The final plan was to give two sequential blasts with a time interval of 3.0 s; the second blast aiming to split the upper structure in half and making the height of the remains lower.

8.5 SUMMARY

The *KI* and its integrated values, as demonstrated in this chapter, can be used to quantitatively verify which columns should be eliminated to best achieve a TC. The numerical results suggest that for a successful demolition, a group of columns with the largest *KI* values should be selected when explosives are ignited in a simultaneous blast, whereas those with the smallest values should be selected when explosives are ignited sequentially, with a final blast set on columns with large index values. Further investigations are ongoing to enable the precise planning of blast intervals and locations required to allow the whole structure to collapse in its own wake, or to lower the height of the remains. These are very important to reduce the risk of damage to neighboring buildings, or human beings who clear up after the demolition, as shown in the last examples.

REFERENCES

Isobe, D., Eguchi, M., Imanishi, K., Sasaki, Z., 2006. Verification of blast demolition problems using numerical and experimental approaches. In: Proc. of the Joint International Conference on Computing and Decision Making in Civil and Building Engineering (ICCCBE'06), pp. 446–455.

Isobe, D., 2014. An analysis code and a planning tool based on a key element index for controlled explosive demolition. International Journal of High-Rise Buildings 3 (4), 243–254.

Kinoshita, M., Hasegawa, A., Matsuoka, S., Nakagawa, K., 1989. An experimental study on explosive demolition methods of reinforced concrete building. Proceedings of the Japan Society of Civil Engineers 403 (VI-10), 173–182.

Williams, G.T., 1990. Explosive demolition of tall buildings in inner city areas. Municipal Engineer 7 (4), 163–173.

Yarimer, E., 1989. Demolition by Controlled Explosion as a Dynamical Process. Structures under Shock and Impact, pp. 411–416.

CHAPTER NINE

Seismic Pounding Analysis of Adjacent Buildings

9.1 INTRODUCTION

In the 1985 Mexican earthquake, many apartment buildings in Mexico City collapsed due to long-period ground motion (Universidad Nacional Autonoma de Mexico, 1985; Ciudad de Mexico, 1986). The city suffered a lot of damage although the epicenter was approximately 400 km away, as shown in Fig. 9.1. Among the collapsed buildings was an apartment called the Nuevo Leon, shown in Fig. 9.2, built in the Tlatelolco district, which had three similar 14-story buildings built next to each other with very narrow gaps connected with expansion joints. But during the earthquake, two buildings among them, the north and the center, collapsed completely as shown in Fig. 9.3. The damage was caused by the impact of the adjacent buildings, which resulted from the change in the natural periods of the buildings from the prior reduction of strength and soil subsidence. An additional effect of the resonance phenomena was caused by long-period ground motion. In Mexico City, extremely soft soil, such as the clay of Lake Texcoco, lies under most parts of the city. This unique subsurface condition resulting from the historical lake bed has distinct resonant low frequencies of approximately 0.5 Hz (Celebi et al., 1987). Therefore, nearly all of the 14-story buildings in the district, which had natural periods of approximately 2 s, were destroyed during the earthquake, as shown in the statistics (Fig. 9.4).

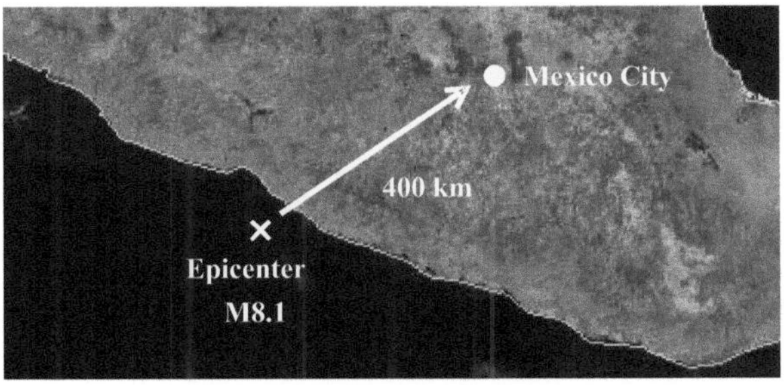

Figure 9.1 Epicenter of the 1985 Mexican earthquake.

Progressive Collapse Analysis of Structures
ISBN: 978-0-12-812975-3
http://dx.doi.org/10.1016/B978-0-12-812975-3.00009-8

Figure 9.2 The Nuevo Leon buildings before the earthquake (Isobe et al., 2012).

Figure 9.3 Collapse of the Nuevo Leon buildings (south building at the far side). *From Isobe, D., 2017. Simulating collapse behaviors of buildings and motion behaviors of indoor components during earthquakes. In: Zouaghi, T. (Ed.), Earthquakes – Tectonics, Hazard and Risk Mitigation. InTech, http://dx.doi.org/10.5772/65490. Available from: http://www.intechopen.com/books/earthquakes-tectonics-hazard-and-risk-mitigation/simulating-collapse-behaviors-of-buildings-and-motion-behaviors-of-indoor-components-during-earthqua.*

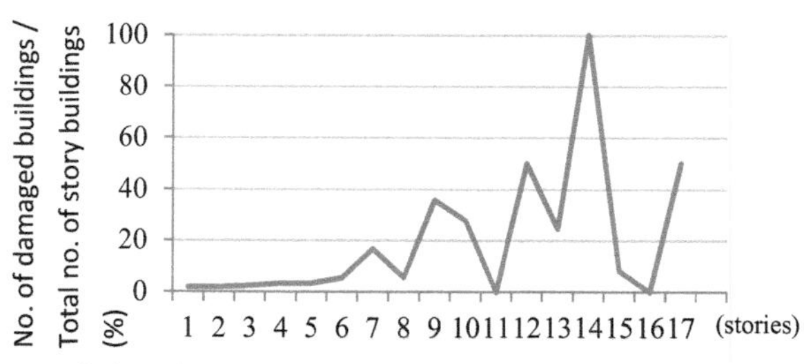

Figure 9.4 Ratio of the damaged buildings versus number of stories in the 1985 Mexican earthquake.

The seismic pounding phenomena due to long-period ground motion were investigated by conducting analyses on two adjacent framed structures with different heights and a simulated model of Nuevo Leon buildings (Isobe et al., 2012; Isobe, 2017). The finite element code based on the ASI-Gauss technique with other related algorithms, described in Chapters 2 and 3, was applied to the analyses. Particularly, the contact release and recontact algorithms described in Chapter 4 were implemented in the code to understand the complex behaviors of structural members during the seismic pounding and collapse sequence. In the analysis of the Nuevo Leon buildings, the natural period of one building was set to be 25% longer than those of the other buildings, as a difference in natural periods was actually observed in similar buildings based on the damage caused by previous earthquakes.

9.2 SEISMIC POUNDING ANALYSIS OF ADJACENT FRAMED STRUCTURES WITH DIFFERENT HEIGHTS

In this section, a seismic pounding analysis is performed on adjacent framed structures to investigate the effects of collisions between them during a long-period ground motion. As shown in Fig. 9.5, simple numerical models of two adjacent framed structures with different heights were constructed to investigate the seismic pounding behavior. One of the framed structures was 12-stories high and the other was 7-stories high. Each story

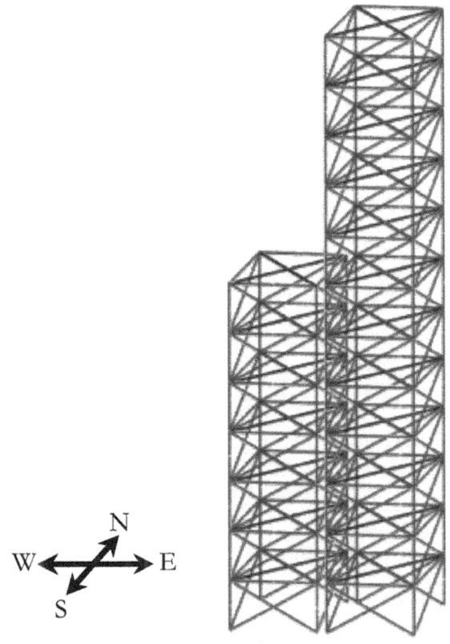

Figure 9.5 Numerical models of the two adjacent framed structures.

was 3.46 m high, with a span length of 6.3 m and a depth of 12.4 m. The distance between the two models was 30 cm. The sectional and the material properties of the models are shown in Tables 9.1 and 9.2, respectively. The floor loads were set to 4.5 kN/m^2. A seismic wave recorded at the transport and communication ministry (SCT) station in Mexico City during the 1985 Mexican earthquake, as shown in Fig. 9.6, was used for the input ground motion. The time increment for the analysis was 1 ms, and the total number of steps was 183,501. The critical curvatures for fracture were set to 3.333×10^{-4}, the critical shear strains to 2.600×10^{-3} and the critical axial tensile strain to 0.17.

No impact occurred between the two structures throughout the analysis when the buildings were both 12-stories high. On the other hand, the models collided during the ground motion when one building was seven-

Table 9.1 Sectional Properties of the Structural Members

	Columns		
	(1–5 F)	(6–10 F)	(11–12 F)
Section (mm)	Box-330 × 330 × 10	Box-280 × 280 × 9	Box-230 × 230 × 7
	Beams	**Floor Slabs**	
Section (mm)	H-292 × 730.0 × 16.2 × 11.6	Box-230 × 230 × 7	

Table 9.2 Material Properties of the Structural Members

	Column	Beam	Floor Slab	Wall Brace
Yield strength (MPa)	3.25×10^2	3.25×10^2	3.25×10^2	2.35×10^2
Young's modulus (GPa)	2.06×10^5	2.06×10^5	2.06×10^5	2.06×10^5
Density (kg/mm^3)	7.90×10^{-6}	7.90×10^{-6}	7.90×10^{-6}	7.90×10^{-6}
Poisson's ratio	0.30	0.30	0.30	0.30

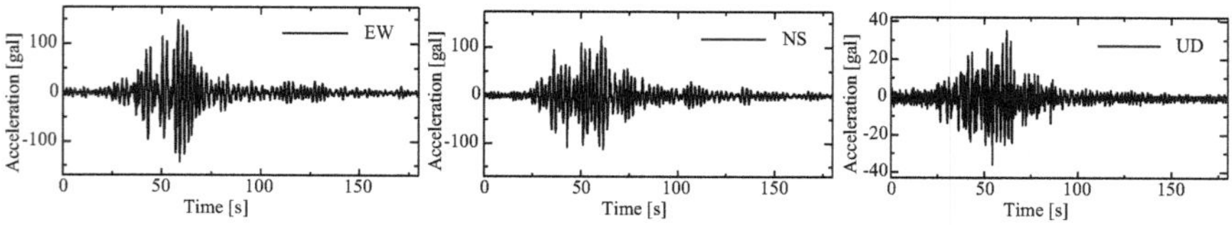

Figure 9.6 Input ground acceleration (SCT wave).

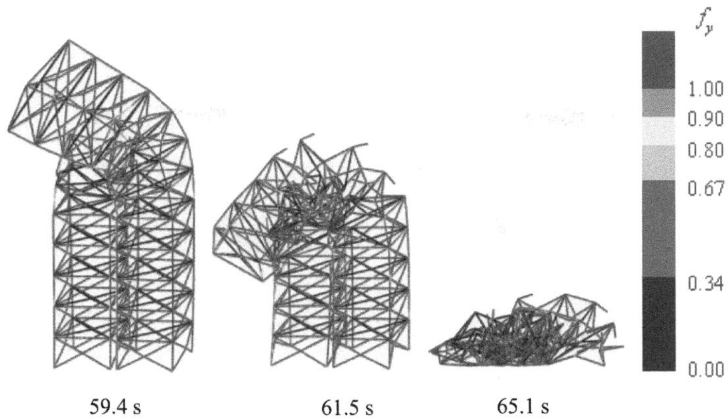

Figure 9.7 Collapse modes of the two adjacent framed structures.

stories high, and both eventually collapsed as shown in Fig. 9.7. The colors represent the distribution of the yield function values f_y defined in Eq. (3.4). This result shows that the distances between adjacent buildings are crucial and that the distances must be sufficiently secured, particularly if the natural periods of the buildings are different.

9.3 SEISMIC POUNDING ANALYSIS OF THE NUEVO LEON BUILDINGS

As shown in Fig. 9.8, a simulated model of the Nuevo Leon buildings with three similar 14-story buildings built consecutively with narrow gaps of 10 cm, was constructed. The model was 42.02 m high and 12.4 m wide, with a total length of 160 m. By referring to the design guidelines of Mexico in 1985, the base shear coefficient was set to 0.06, and the axial force ratio on

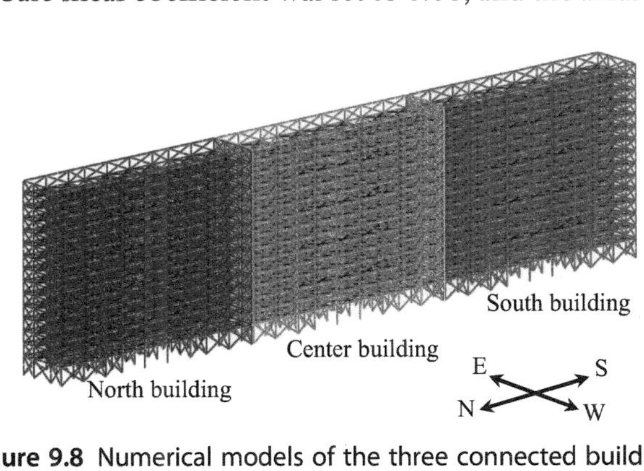

Figure 9.8 Numerical models of the three connected buildings.

the first floor was set approximately to 0.5. The dead load for each floor was 4.0 kN/m^2 and the damping ratio was 5%. The Nuevo Leon buildings were originally built with reinforced concrete (RC) members; however, the model was intentionally constructed with steel members to easily verify the influence of the structural parameters, such as member fracture strains. The critical curvatures for the fracture were set to 3.333×10^{-4}, the critical shear strains to 3.380×10^{-4}, and the critical axial tensile strain to 0.17. The critical shear strains used were lower than the strain values of the steel members to consider the characteristics of RC beams. The sectional properties of the structural members are shown in Table 9.3. The time increment was set to 1 ms, and the calculation was performed for 90,000 steps. The analysis took approximately 4 days using a standard personal computer (CPU: 2.93 GHz Xeon).

As shown in Table 9.4, the natural period of the north building model was set to be 25% longer than the other periods by lowering the structural strengths of the columns. The difference of natural periods was actually observed in similar buildings built near the site (see Table 9.5), caused by the damage from previous earthquakes (Ciudad de Mexico, 1986). The EW, NS, and UD components of the SCT seismic wave shown in Fig. 9.6 were subjected to the fixed points on the ground floor. As mentioned earlier, the intensity period of the seismic wave was approximately 2 s because of the reclaimed soft soil of Mexico Valley. As can be seen in Fig. 9.9, the collisions

Table 9.3 Sectional Properties of the Structural Members

	Columns		
	(1–5 F)	**(6–10 F)**	**(11–14 F)**
Section (mm)	Box-330 × 330 × 10	Box-280 × 280 × 9	Box-230 × 230 × 7
	Beams	**Floor Slabs**	
Section (mm)	H-292 × 730.0 × 16.2 × 11.6	Box-230 × 230 × 7	

Table 9.4 Natural Period of Each Building Model

	NS (s)	EW (s)
North	1.5	1.72
Center	1.2	1.65
South	1.2	1.65

Table 9.5 Natural Period of Chihuanua, a Similar Building

	NS			EW		
Building No.	**1**	**2**	**3**	**1**	**2**	**3**
Natural period (s)	1.39	1.11	1.13	1.94	1.63	1.77
Ratio of period to No.2 building	1.25	1.0	1.02	1.19	1.0	1.09

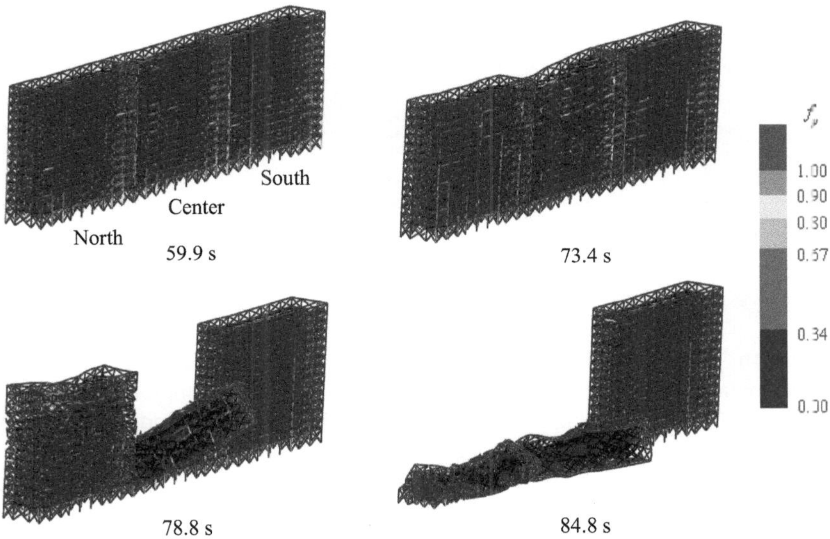

Figure 9.9 The collapse behaviors of the simulated model of the Nuevo Leon buildings under long-period ground motion.

of the buildings are triggered by the difference in the natural periods between adjacent buildings. If we look closely to the details as shown in Fig. 9.10, first a collision occurs between the north and center buildings due to the difference of the natural periods, and the plastic region spreads through the beams and columns. Then the columns near the impact point occasionally lose their structural strengths resulting from the continuous pounding sequence. The collapse of the center building is initiated at the ceiling of the ninth floor because of the continuous collisions from both sides, which begins approximately 70 s from the start of seismic activity. Although the north building collapses a few seconds after the center, the south building withstands the collisions and does not collapse as shown in Fig. 9.9.

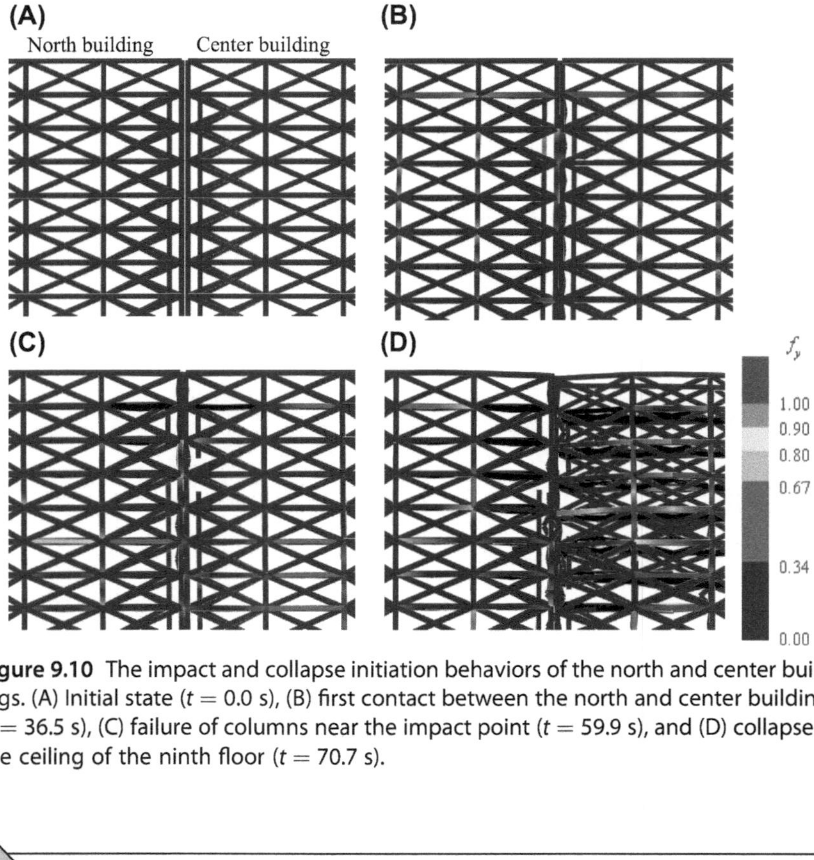

Figure 9.10 The impact and collapse initiation behaviors of the north and center buildings. (A) Initial state ($t = 0.0$ s), (B) first contact between the north and center buildings ($t = 36.5$ s), (C) failure of columns near the impact point ($t = 59.9$ s), and (D) collapse of the ceiling of the ninth floor ($t = 70.7$ s).

9.4 SUMMARY

The numerical results shown in this chapter clarify the possibility that long-period ground motion may cause extra damage in high-rise architectures due to inter-building collisions, if the distances between them are not sufficiently secured. Extra caution may be required if the natural periods of the adjacent buildings are different, which can easily occur, for example, if their heights are different.

REFERENCES

Celebi, M., Pince, J., Dietel, C., Onate, M., Chavez, G., 1987. The culprit in Mexico City — amplification of motions. Earthquake Spectra 3 (2), 315—328.

Ciudad de Mexico, 1986. Programa de reconstruccion Nonoalco/Tlatelolco, Tercera reunion de la Comision Tecnica Asesora (The Third meeting of the Technical Commission Advises).

Isobe, D., 2017. Simulating collapse behaviors of buildings and motion behaviors of indoor components during earthquakes. In: Zouaghi, T. (Ed.), Earthquakes – Tectonics, Hazard and Risk Mitigation. InTech, ISBN 978-953-51-2885-4, pp. 341–363.

Isobe, D., Ohta, T., Inoue, T., Matsueda, F., 2012. Seismic pounding and collapse behavior of neighboring buildings with different natural periods. Natural Science 4 (8A), 686–693.

Universidad Nacional Autonoma de Mexico, 1985. The Earthquake of September 19th, 1985, Inform and Preliminary Evaluation.

CHAPTER TEN

Seismic Collapse Analysis of the CTV Building

10.1 INTRODUCTION

A strong earthquake of magnitude 6.3 struck the city of Christchurch, New Zealand, on February 22, 2011. It was an aftershock of the September 4, 2010, Darfield earthquake and was named as the 2011 Lyttelton aftershock. A six-story reinforced-concrete building known as the Canterbury Television (CTV) building (Fig. 10.1) collapsed in the 2011 aftershock with the exception of north-wall complex left standing (Fig. 10.2); the accident killed 115 people. An investigation report (Hyland and Smith, 2012) by the Department of Building and Housing, New Zealand, estimated the collapse sequence of the building, as shown in Fig. 10.3, based on the narratives from the witnesses. It said that the entire building somehow vibrated with a twist mode, which took away the capacity of the east-side column at Level 4 and caused an eastward tilt in the upper levels. The other columns could not support the redistributed floor loads and began to fail progressively. Consequently, the entire building collapsed with all floors lying on top of each

Figure 10.1 CTV building before the earthquake. *Photo by Phillip Pearson, from https://en.wikipedia.org/wiki/CTV_Building#/media/File:Canterbury_Television_building,_2004_crop.jpg.*

Progressive Collapse Analysis of Structures
ISBN: 978-0-12-812975-3
http://dx.doi.org/10.1016/B978-0-12-812975-3.00010-4

Figure 10.2 Ruins of the CTV building after the earthquake. *From https://en.wikipedia.org/wiki/CTV_Building#/media/File:Ruins_of_the_Canterbury_Television_(CTV)_building,_24_February_2011.jpg.*

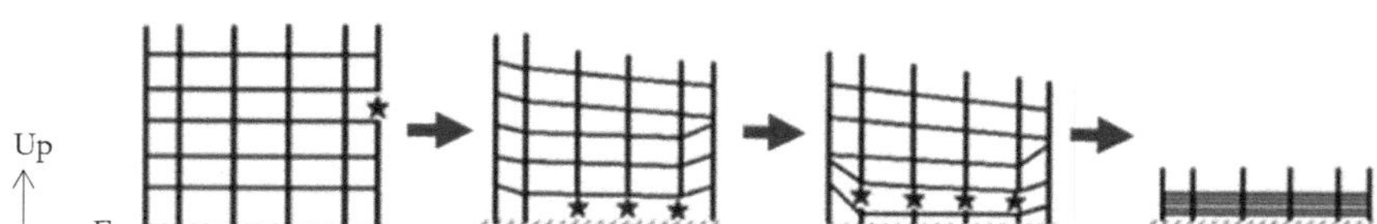

Figure 10.3 Possible collapse sequence from the south front (Hyland and Smith, 2012).

other. Some numerical results of nonlinear seismic analyses (Bradley and Stuart, 2012) were reported shortly after, which contained numerous elasto-plastic analysis results of the CTV building under different seismic waves of the 2010 Darfield earthquake and 2011 Lyttelton aftershock. The final report (Canterbury Earthquakes Royal Commission, 2012) stated that one of the reasons of the collapse was the failure of the structural designer to consider load tracking through the beam—column joints. The joints lacked ductility and were brittle in character. It was also reported that it only took about 10—20 s from the start of the seismic activity until the total collapse of the building.

According to these reports, there were two main possible factors of the collapse. One factor was the original structural design of the building. The building might not have maintained the necessary anti-seismic performance because of its unbalanced distribution of anti-seismic walls and insufficient strengths of joints between beams and columns. The second factor was insufficient restoration of damage from previous earthquakes. It was said that the building was not completely restored after the 2010 Darfield earthquake despite damages such as cracks in the columns and walls.

This chapter describes an investigation carried out using the ASI-Gauss code. The CTV building was modeled based on the actual structural design (Hyland and Smith, 2012) and a pushover analysis was performed. Furthermore, a collapse analysis was conducted to see the response under seismic excitation and collapse sequence of the building. A constitutive equation of the reinforced-concrete member was implemented in the ASI-Gauss code and applied to the analyses.

10.2 CONSTITUTIVE EQUATION OF THE REINFORCED-CONCRETE MEMBERS

In this chapter, a degrading trilinear model with viscous damping (Umemura, 1973; Architectural Institute of Japan, 1997) was used for the constitutive equation of reinforced-concrete members. Its hysteresis curve is shown in Fig. 10.4. Bending and shear deformations were assumed to occur independently, and the constitutive equations shown below were used in this model.

10.2.1 Bending Yield Strength

- Columns

$$M_y = 0.5(a_g\sigma_y + N)g_1D \quad \text{when} \quad N_{\min} \leqq N \leqq 0 \tag{10.1}$$

$$M_y = 0.5a_g\sigma_y g_1 D + 0.5ND\left(1 - \frac{N}{bDF_c}\right) \quad \text{when} \quad 0 \leqq N \leqq N_b \tag{10.2}$$

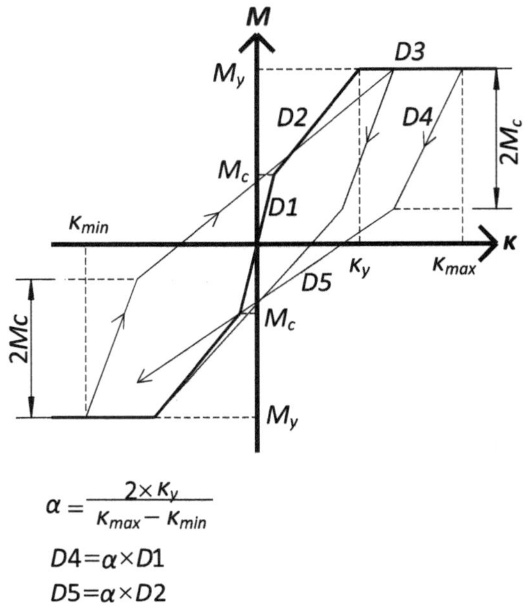

Figure 10.4 Degrading trilinear model.

$$M_y = \{0.5a_g\sigma_y g_1 D + 0.24(1 + g_1)(3.6 - g_1)bD^2F_c\}\frac{N_{\max} - N}{N_{\max} - N_b}$$
$$\text{when } N_b \leqq N \leqq N_{\max} \tag{10.3}$$

where

$$N_{\min} = -a_g\sigma_y \tag{10.4a}$$

$$N_{\max} = bDF_c + a_g\sigma_y \tag{10.4b}$$

$$N_b = 0.22(1 + g_1)bDF_c \tag{10.4c}$$

- Beams

$$M_y = 0.9a_t\sigma_y d \tag{10.5}$$

10.2.2 Bending Crack Strength

$$M_c = \frac{1}{3}M_y \tag{10.6}$$

10.2.3 Shear Ultimate Strength

$$Q_y = \left\{ \frac{0.115 k_u k_p (180 + F_c)}{M/Qd + 0.115} + 2.7\sqrt{p_w \sigma_{wy}} + 0.1 \frac{N}{bD} \right\} bj \tag{10.7}$$

10.2.4 Shear Crack Strength

$$Q_c = \frac{1}{3} Q_y \tag{10.8}$$

10.2.5 Stiffness Reduction Ratio at the Yield Point

$$\alpha_y = \left(0.043 + 1.65 n p_t + 0.043 \frac{M}{Qd} + 0.33 \eta_0 \right) \left(\frac{d}{D} \right)^2 \tag{10.9}$$

where b is the column width, D is the column height, h_0 is the inner measured height of the column, d is the equivalent height $= 0.9D$, g_1 is the distance among the centers of gravity of the reinforcements, F_c is the compressive strength of concrete, σ_y is the yield stress of the tension reinforcement, σ_{wy} is the tension yield stress of the shear reinforcement, n is the ratio of Young's moduli of the concrete and reinforcement, N is the normal force, M/Qd is the shear span ratio $= h_0/2d$, η_0 is the normal force ratio $= N/bDF_c$, j is the distance between the centers of stress $= 7d/8$, k_u is the correction factor of member dimension $= 0.7$, k_p is the correction factor of the tension reinforcement ratio $= 0.82 p_t^{0.23}$, p_t is the tension reinforcement ratio $= a_t/bD$, p_w is the shear reinforcement ratio $= a_w/b\chi$, a_t is the cross-sectional area of the tension reinforcement, a_g is the cross-sectional area of all reinforcements, and χ is the distance between each shear reinforcement.

The strength failure of the reinforced-concrete members was determined by evaluating the shear strains and tensional normal strain for the elements that approach the bending yield strength or shear ultimate strength. The following equations were used to determine the member strength failure.

$$\left| \frac{\gamma_{xz}}{\gamma_{xz0}} \right| - 1 \geq 0 \quad \text{or} \quad \left| \frac{\gamma_{yz}}{\gamma_{yz0}} \right| - 1 \geq 0 \quad \text{or} \quad \left(\frac{\varepsilon_z}{\varepsilon_{z0}} \right) - 1 \geq 0 \tag{10.10}$$

where γ_{xz}, γ_{yz}, ε_z, and γ_{xz0}, γ_{yz0}, ε_{z0} are the shear strains along the x- and y-axes, axial tensile strain, and critical values for these strains, respectively.

10.3 NUMERICAL MODEL

A CTV building model constructed based on the structural design in the investigation report (Hyland and Smith, 2012) is shown in Fig. 10.5. The location plan of the columns, beams, and anti-seismic walls is shown in Fig. 10.6. It was a six-story reinforced-concrete building with an elevator shaft at the north part of the building, which was reinforced with anti-seismic walls. There were no beams except the perimeter area, which ran southward. The sectional sizes and dimensions of the columns and beams are listed in Tables 10.1 and 10.2, respectively. The description "H" in the tables indicates a deformed bar and "R" indicates a round steel bar. The yield strengths of the deformed bar and round steel bar are 380 and 275 MPa, respectively. The compressive strengths of concrete for

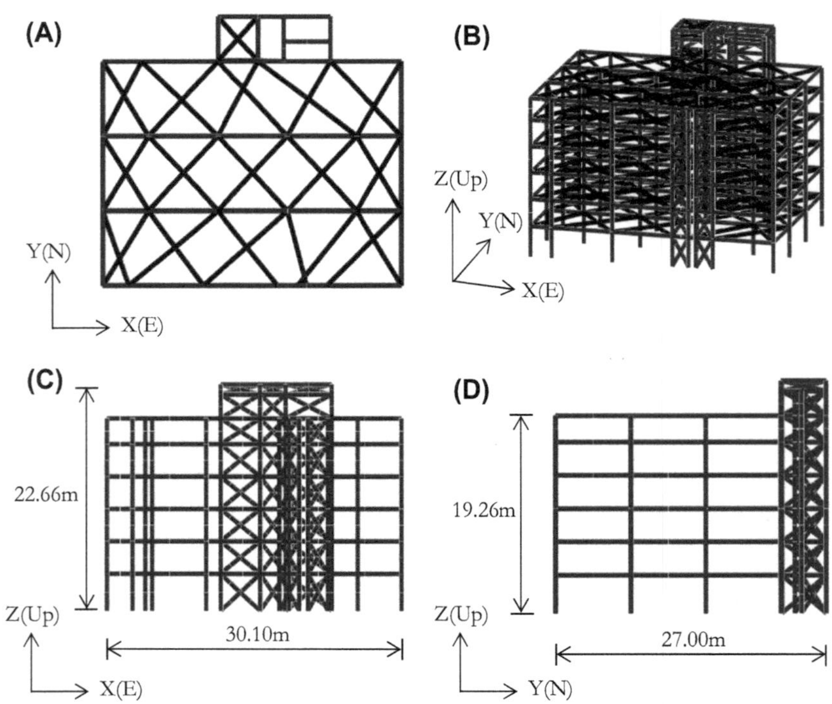

Figure 10.5 Numerical model of the CTV building. (A) Plan, (B) bird view, (C) front view, and (D) side view.

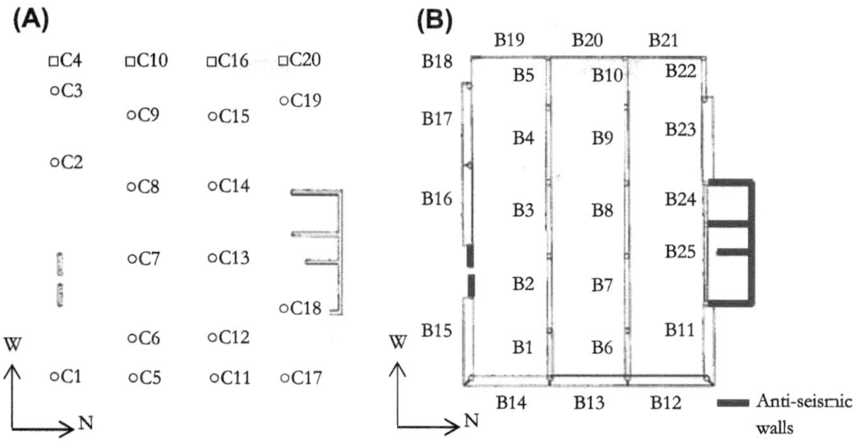

Figure 10.6 Location plan of the structural members. (A) Columns and (B) beams and anti-seismic walls.

Table 10.1 Sectional Sizes of the Columns

Name	Levels	Sectional Sizes (mm)	Re-bars
C1	1–6	354.5 × 354.5	6-H20
C2	↓	↓	R12@250
C3			↓
C5			
C6			
C7			
C8			
C9			
C11			
C12			
C13			
C14			
C15			
C17			
C19			
C4	1–6	300 × 400	4-H20
C10	↓	↓	R10@250
C16			↓
C20			
	1	400 × 400	6-H20
C18			R6@250
	2–8	354.5 × 354.5	6-H20
			R6@250

Table 10.2 Sectional Sizes of the Beams

Name	Levels	Sectional Sizes (mm)	Re-bars
B1	1–6	400 × 350	3-H28 R12@125
B5	↓	↓	↓
B6	↓	↓	↓
B10	↓	↓	↓
B2	1–6	400 × 550	8-H28 R12@200
B3	↓	↓	↓
B4	↓	↓	↓
B7	↓	↓	↓
B8	↓	↓	↓
B9	↓	↓	↓
B11	1–6	550 × 960	8-H24 R12@200
B12	↓	↓	↓
B13	↓	↓	↓
B14	↓	↓	↓
B15	↓	↓	↓
B16	↓	↓	↓
B17	↓	↓	↓
B23	↓	↓	↓
B18	1–6	400 × 550	1-H20 2-H24 R12@200
B22	↓	↓	↓
B19	1–6	140 × 430	1-H20 R10@200
B20	↓	↓	↓
B21	↓	↓	↓
B24	1–6	400 × 550	4-H24 R6@300
B25	1–6	400 × 550	8-H24 R12@200

the Levels 1 and 2 columns are 35 and 30 MPa, respectively, and that for the Level 3 and upper-level columns is 25 MPa. The columns with a round section (C1, C2, C3, C5, C6, C7, C8, C9, C11, C12, C13, C14, C15, C17, C18, and C19 in Fig. 10.6; Tables 10.1 and 10.2) are replaced with regular-square-section columns of identical sectional area, as shown in Table 10.3, to coincide with the constitutive equation of the reinforced-concrete members. The north-wall complex and anti-seismic walls at the south face are modeled with braces of equivalent strengths. Joint elements with a length of 0.1 m are introduced between the main building and the north-wall complex. The floor load was set to

Table 10.3 Original and Substituted Reinforced-Concrete Members

	Original Member	Substituted Member
Section	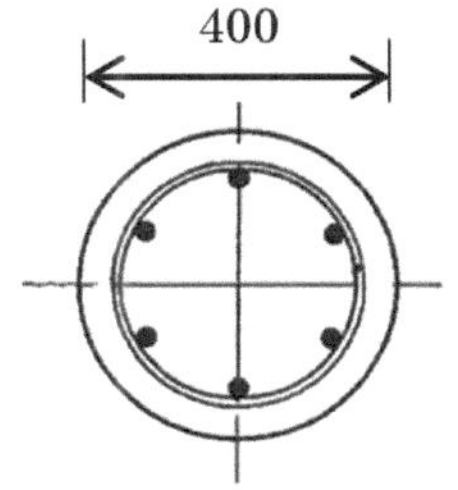	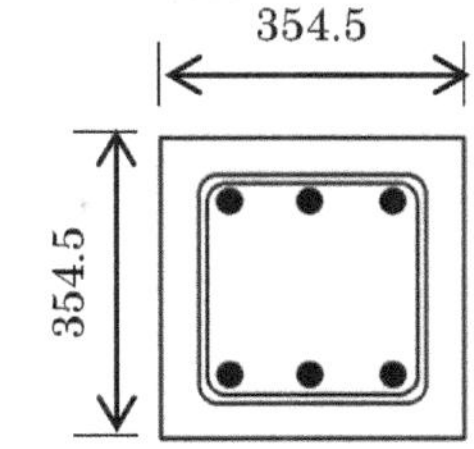
Re-bar	6-H20 R6@250	6-H20 R6@250

Table 10.4 Critical Strains for Member Strength Failure

Axial tensile strain ε_{z0}	0.17
Shear strain along x-axis γ_{xz0}	0.20
Shear strain along y-axis γ_{yz0}	0.20

400 kgf/m^2 and distributed to the floors in the model by increasing the member density of beams and floor members. The critical strains for the member strength failure were selected based on the evaluation results of anti-seismic performance of reinforced-concrete pillars with high-strength reinforcement bars, as shown in Table 10.4 (Kitamura et al., 2012). As we applied the ASI-Gauss code, all structural members were modeled with two linear Timoshenko beam elements per member. The natural periods of the numerical model in the EW and NS directions were 0.98 and 1.15 s (1.0 and 1.3 s in (Bradley and Stuart, 2012)), respectively. A damping ratio of 2.5% was used. The total number of elements was 2238, and the total number of nodes was 1540.

10.4 PUSHOVER ANALYSIS OF THE CTV BUILDING

First, a pushover analysis of the CTV building was performed. Incremental loads based on the A_i distribution were applied in the east, west, south, and north directions to observe the differences in static responses around the north-wall complex. The base shear force–displacement

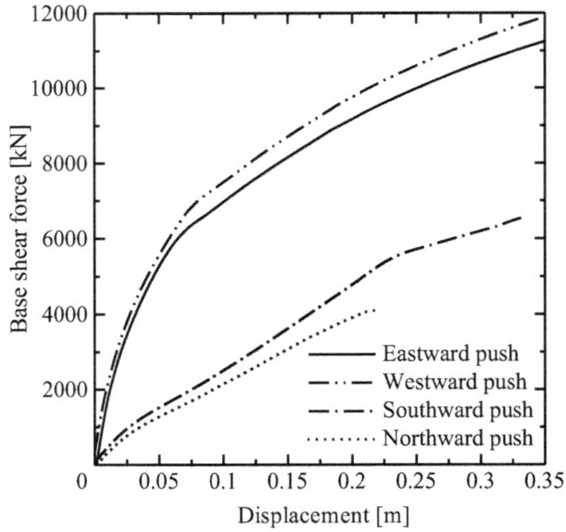

Figure 10.7 Base shear force—displacement relation.

relation at Level 1 is shown in Fig. 10.7 for each direction. The behaviors have identical tendencies in the eastward and westward cases, and prove that these cases have almost identical strengths. The behaviors in the southward and northward cases also tend to show the identical tendencies with almost identical strengths. However, the strength of the building in the southward (or northward) direction is almost half of that in the eastward (or westward) direction, and there is a strong unbalance between these directions. It is because no beam ran southward except in the perimeter area, and anti-seismic walls were excessively distributed in the eastward direction.

10.5 COLLAPSE ANALYSIS OF THE CTV BUILDING UNDER THE 2011 LYTTELTON AFTERSHOCK

Next, a collapse analysis of the CTV building was performed using the ASI-Gauss code. A Christchurch Resthaven (REHS) wave observed near the site during the 2011 Lyttelton aftershock was used as an input wave. The acceleration data of the REHS wave are shown in Fig. 10.8, and its acceleration response spectrum is shown in Fig. 10.9. The acceleration data in three directions reached their peak levels at 10—13 s from the beginning of the seismic activity. The predominant period of the input wave was 1.0 s in the EW direction, which almost matched the natural period of the model

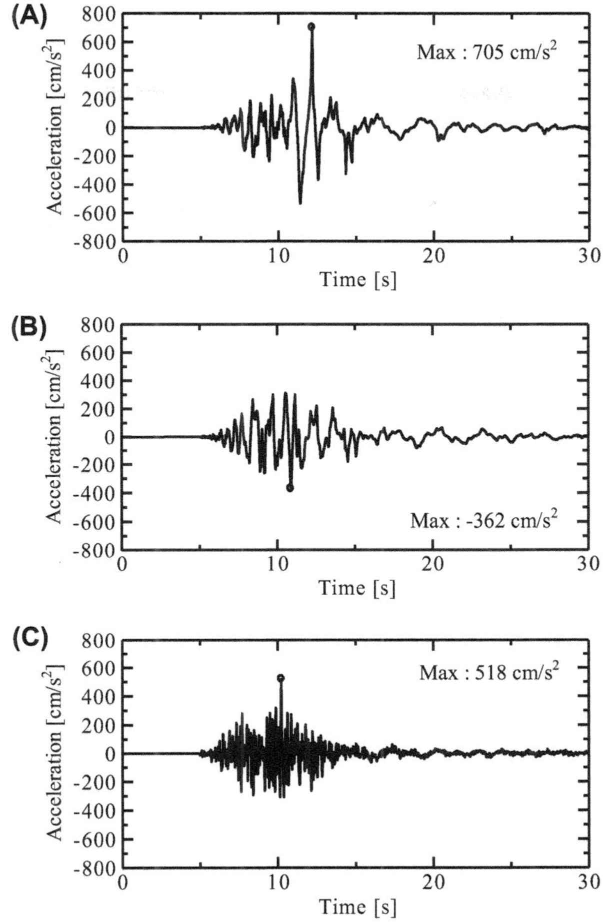

Figure 10.8 Input wave (REHS wave). (A) EW, (B) NS, and (C) UD.

(0.98 s) in this direction, and 0.5 s in the NS direction. The total time duration in the analysis was 30 s, and the incremental time was set to 1.0 ms.

The bird-view figures of the building behavior obtained from the analysis are shown in Fig. 10.10, and the front-view figures at the same elapsed times are shown in Fig. 10.11. The colors in the figures indicate the yield function values, which were evaluated, in this analysis, using the following equation:

$$f_y = \left(\frac{M_x}{M_{x0}}\right)^2 + \left(\frac{M_y}{M_{y0}}\right)^2 + \left(\frac{V_x}{V_{x0}}\right)^2 + \left(\frac{V_y}{V_{y0}}\right)^2 \tag{10.11}$$

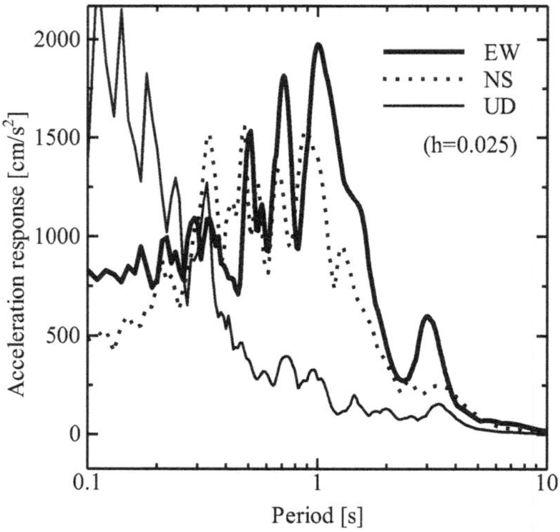

Figure 10.9 Acceleration response spectrum (REHS wave).

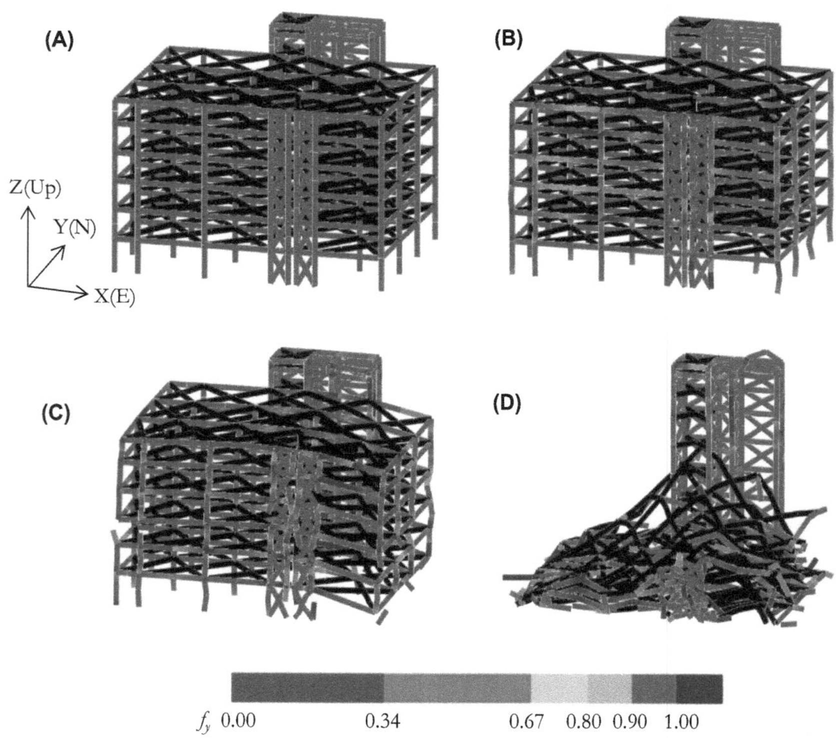

Figure 10.10 Collapse behavior of the CTV building (bird view). (A) 0.0 s, (B) 13.7 s, (C) 15.5 s, and (D) 25.0 s.

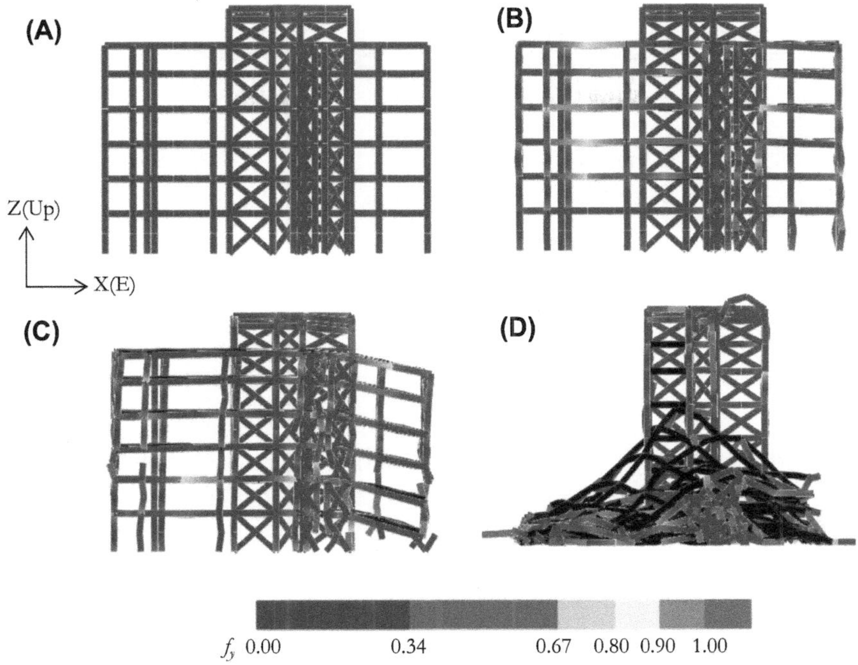

Figure 10.11 Collapse behavior of the CTV building (front view). (A) 0.0 s, (B) 13.7 s, (C) 15.5 s, and (D) 25.0 s.

where f_y is the yield function, M_x and M_y are the bending moments around the x- and y-axes, and V_x and V_y are the shear forces along the x- and y-axes, respectively. The terms with the subscript 0 are values that result in a fully yielded section in an element if they independently act on the cross section. The elements that were determined to be in the member strength failure phase are omitted from the graphics. Member strength failure of elements can be observed (the elements are omitted from the graphics because of the determination) at the east-side column on Level 3 at 13.7 s in the figures. The entire building significantly tilted to the east and totally collapsed with all floors lying on top of each other, with only the north-wall complex left standing. There was a twist mode vibration around the north-wall complex observed before the column on Level 3 first lost its member strength. The sequence took less than 20 s from the start of the seismic activity till the total collapse. Shear deformations that occurred at the columns may well have caused the outcome, because all columns that lost their member strengths were determined based on the shear strain. The behavior of the building was consistent with the general sketch (Fig. 10.3) and the statement in the collapse

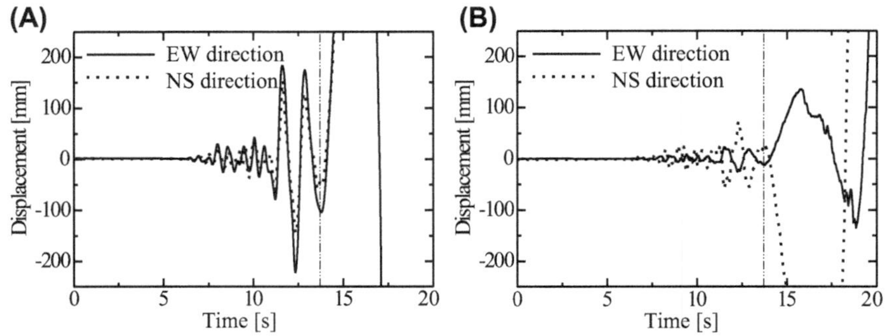

Figure 10.12 Responses of the columns on Level 6. (A) Southeast corner column (C1) and (B) northwest corner column (C20).

investigation report of the CTV building (Hyland and Smith, 2012), which concluded that a twist motion of the entire building may have caused the member strength failure of the columns at the east side on higher-level floors.

Fig. 10.12A shows the responses of the southeast corner column (C1) on Level 6 in the EW and NS directions, and Fig. 10.12B shows those of the northwest corner column (C20) on Level 6. The time when the collapse began is indicated in these figures by a vertical, two-dot chain line. The responses of column C1 in both EW and NS directions were in the same phase and increased from approximately 11 s. However, the responses of column C20 in the EW and NS directions were in the opposite phases, and the amplitudes were small compared to that of column C1. These results indicate the difference in response behavior depending on the locations in the building.

The relative story displacements of column C1 on each floor in the EW and NS directions are shown in Figs. 10.13 and 10.14, respectively.

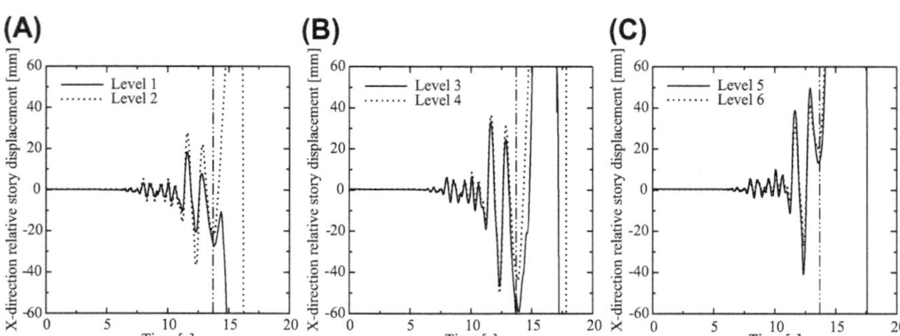

Figure 10.13 Responses of the southeast corner column (C1) in the EW direction. (A) Levels 1, 2, (B) Levels 3, 4, and (C) Levels 5, 6.

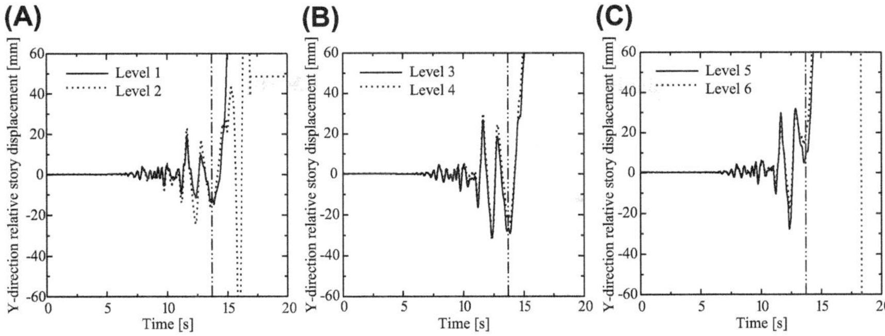

Figure 10.14 Responses of the southeast corner column (C1) in the NS direction. (A) Levels 1, 2, (B) Levels 3, 4, and (C) Levels 5, 6.

In the same manner, the relative story displacements of the southwest corner column (C4), northeast corner column (C17), and C20 are shown in Figs. 10.15–10.20, respectively. Fig. 10.13 indicates that the amplitude of the relative story displacement on Level 3 significantly increased before the collapse began, which may have caused the member strength failure of the columns on this level. It is also noted that the displacements in both the EW and NS directions on Level 3 and upper levels are relatively larger than those on Levels 1 and 2 (Figs. 10.13 and 10.14). Figs. 10.13–10.16 show that the behaviors of the columns at the south face of the building (C1 and C4) follow the same tendency; all relative story displacements in the EW direction are larger than those in the NS direction. Meanwhile, a different tendency is confirmed in the behaviors of the columns at the north face of the building (C17 and C20) as shown in Figs. 10.17–10.20; all displacements in the NS direction are larger than those in the EW

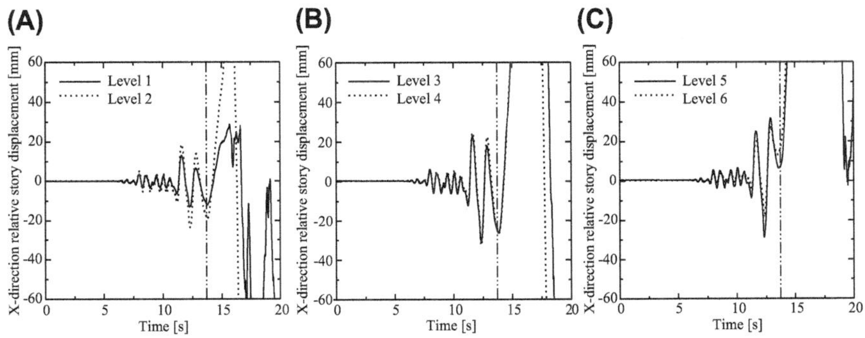

Figure 10.15 Responses of the southwest corner column (C4) in the EW direction. (A) Levels 1, 2, (B) Levels 3, 4, and (C) Levels 5, 6.

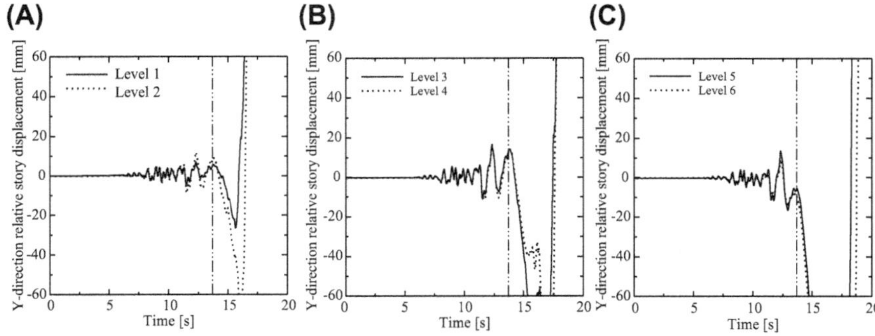

Figure 10.16 Responses of the southwest corner column (C4) in the NS direction. (A) Levels 1, 2, (B) Levels 3, 4, and (C) Levels 5, 6.

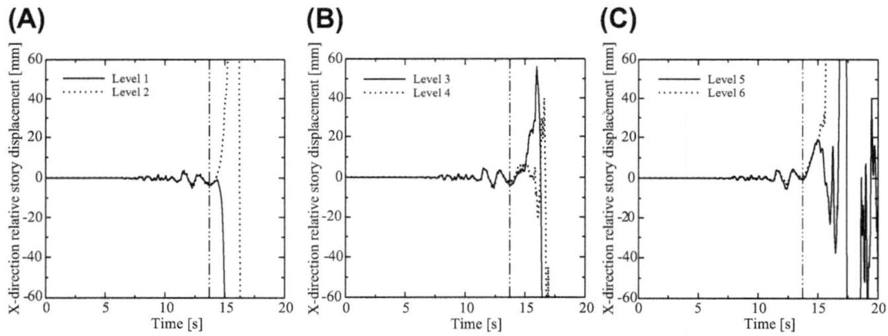

Figure 10.17 Responses of the northeast corner column (C17) in the EW direction. (A) Levels 1, 2, (B) Levels 3, 4, and (C) Levels 5, 6.

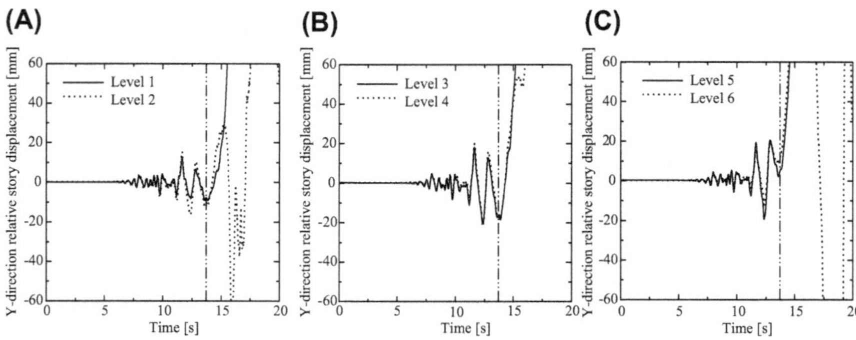

Figure 10.18 Responses of the northeast corner column (C17) in the NS direction. (A) Levels 1, 2, (B) Levels 3, 4, and (C) Levels 5, 6.

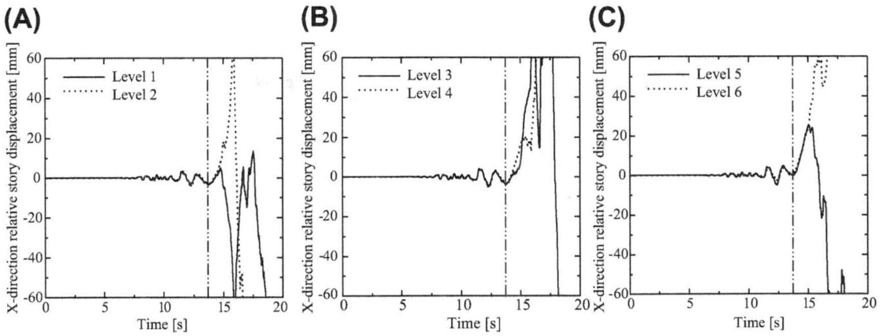

Figure 10.19 Responses of the northwest corner column (C20) in the EW direction. (A) Levels 1, 2, (B) Levels 3, 4, and (C) Levels 5, 6.

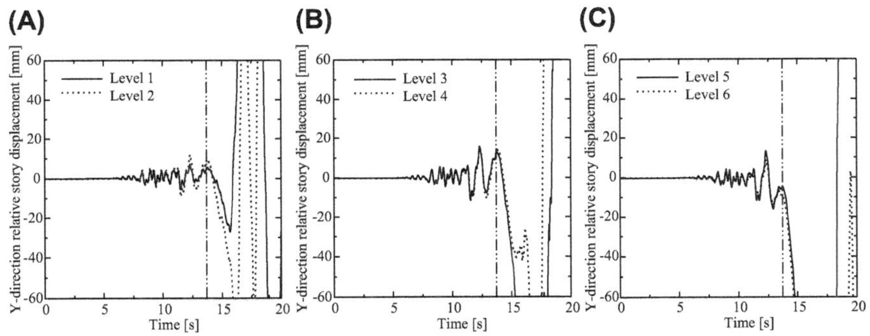

Figure 10.20 Responses of the northwest corner column (C20) in the NS direction. (A) Levels 1, 2, (B) Levels 3, 4, and (C) Levels 5, 6.

direction. In addition, the displacements in both directions are relatively smaller than those of the columns along the south face.

The displacement tracks of the four corner columns (C1, C4, C17, and C20) on Level 6 before the collapse began are plotted in Fig. 10.21. The columns along the south face (C1 and C4) tend to shake significantly in the EW direction, whereas the columns along the north face (C17 and C20) tend to shake mainly in the NS direction. This figure clearly indicates a twist mode vibration of the building around the north-wall complex. Furthermore, column C1 displaced more in the EW direction than column C4. When a sine wave with a period of 1.02 s was applied in the EW direction, C1 was resonant with the input wave and responded more strongly than C4. However, the response of C1 decreased, and that of C4 increased when a sine wave with a period of 1.12 s was applied; C4 was resonant with the input wave in the latter case. These differences occurred because of the slight differences

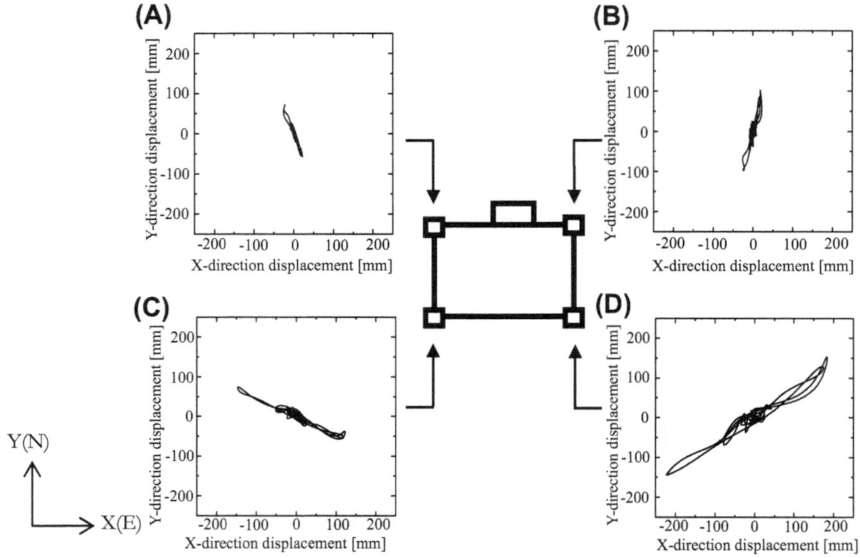

Figure 10.21 Displacement tracks of the four corner columns. (A) Northwest corner (C20), (B) northeast corner (C17), (C) southwest corner (C4), and (D) southeast corner (C1).

in distance from the north-wall complex to each column. The predominant period of the REHS wave in the EW direction was 1.0 s, which coincidentally matched the natural period at the location of C1. This unexpected coincidence may have caused the building to initiate the collapse from the east side.

10.6 SUMMARY

In this chapter, a static pushover analysis and a dynamic collapse analysis were performed on a numerical model of the CTV building, which collapsed during the 2011 Lyttelton aftershock in New Zealand. The collapse sequence of the building was consistent with the summary in the investigation report. It can be concluded from the numerical results that the main causes of the twist mode vibration around the north-wall complex, which occurred before the collapse started, were the absence of southward beams and an unbalanced distribution of anti-seismic walls. The period of the twist mode vibration at the southeast corner of the building coincidentally matched the predominant period of the seismic wave in EW direction, which might had led to the deterioration of the columns. After the collapse began, the columns with insufficient strengths could not support the redistributed floor loads and began to fail progressively. Consequently, the entire building collapsed with all floors lying on top of each other. It took less than 20 s from the start of the seismic activity until the total collapse. Furthermore, a continuous analysis applying a 2010 Darfield seismic wave followed by the 2011 Lyttelton wave was conducted; nevertheless, the results did not show any difference. As mentioned in the investigation reports, the insufficient restoration of damage due to previous earthquakes did not appear to affect the result significantly.

REFERENCES

Architectural Institute of Japan, 1997. Design Guidelines for Earthquake Resistant Reinforced Concrete Buildings Based on Inelastic Displacement Concept (Draft) (in Japanese).

Bradley, D., Stuart, T., 2012. Non-Linear Seismic Analysis Report. Compusoft Engineering Limited, New Zealand.

Canterbury Earthquakes Royal Commission, 2012. Final Report – Part Three (Volume 6: Canterbury Television Building (CTV)). http://canterbury.royalcommission.govt.nz/Commission-Reports.

Hyland, C., Smith, A., 2012. CTV Building Collapse Investigation. Department of Building and Housing, New Zealand.

Kitamura, T., Yoshikawa, T., Tamakoshi, T., 2012. Evaluation of anti-seismic performance of road bridge RC pillars using high-strength reinforcement bars. In: Proc. Workshop on Land and Infrastructure Technology, National Institute for Land and Infrastructure Management, Ministry of Land, Infrastructure and Transport (in Japanese).

Umemura, H., 1973. Dynamic seismic design for reinforced concrete buildings. Giho-do (in Japanese).

CHAPTER ELEVEN

Debris-Impact Analysis of Steel-Framed Building in Tsunami

11.1 INTRODUCTION

The huge tsunami after the Great East-Japan Earthquake in 2011 was reported to have caused additional damages to the buildings by washing up large debris such as ships on land (NILIM, 2011; PARI, 2011). In this chapter, a steel-framed building subjected to seismic excitation, tsunami, and debris collision is analyzed in a single continuous simulation to see the effects of each force on the collapse behavior of the building (Isobe, 2017). A seismic excitation recorded in Kesennuma-shi was first applied to the model, followed by an input of the drag force and buoyant force due to tsunami wave. At the last phase of the analysis, a debris model with a velocity was collided to the building and the collapse behavior of the building was analyzed.

11.2 NUMERICAL MODEL AND CONDITIONS

As shown in Fig. 11.1, a numerical model of a six-story, three-span, steel-framed building, with a floor height of 3.6 m and a span length of

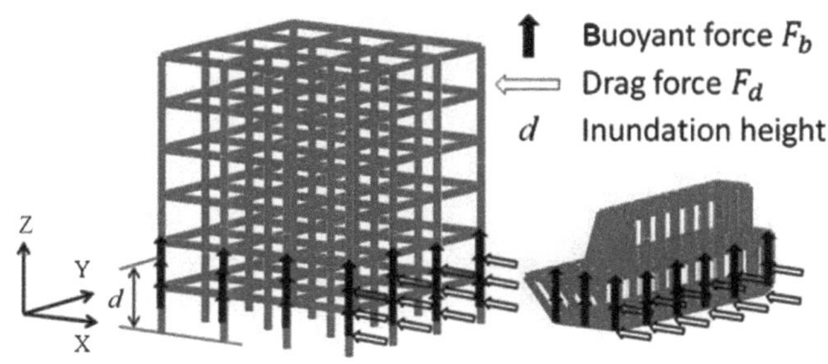

Figure 11.1 Drag forces and buoyant forces applied to the building and tsunami debris.

Progressive Collapse Analysis of Structures
ISBN: 978-0-12-812975-3
http://dx.doi.org/10.1016/B978-0-12-812975-3.00011-6

6 m, was constructed. The base shear coefficient of the building was set to 0.3 and a floor load of 400 kgf/m^2 was applied to all floors. The debris model is a ship made up of aluminum alloy which weighs 110 tons with a length of 27 m, a width of 6 m, and a height of 8 m. First of all, a Kesennuma wave shown in Fig. 11.2, observed in the Great East-Japan Earthquake, was applied only to the building as an input for 150 s from the start of the seismic activity.

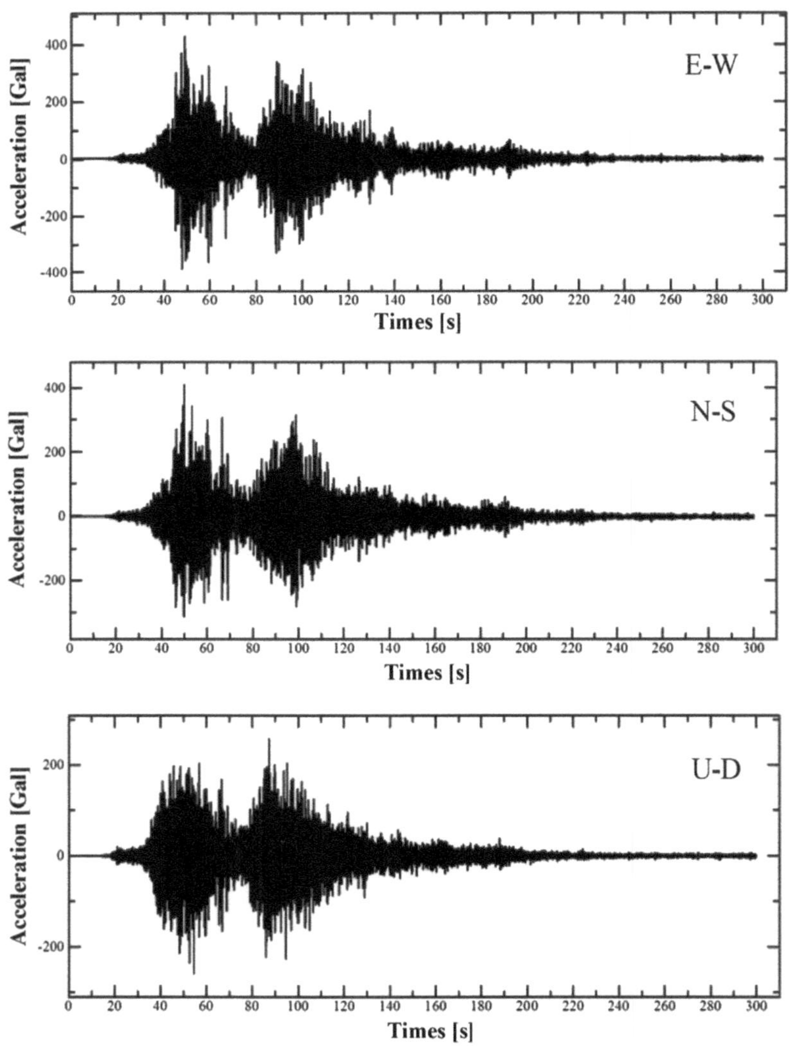

Figure 11.2 Input ground acceleration (Kesennuma 100% wave).

Then, buoyant forces were applied statically, and drag forces were applied dynamically, at the locations indicated in Fig. 11.1, for the next second. The buoyant force applied to the debris balances with the weight and can be expressed as

$$F_b = \rho_s g V \tag{11.1}$$

where ρ_s is the density of seawater including debris which is 1200 kg/m^3, g the gravitational acceleration, and V the volume of water removed by an object. In case of the building model with walls under the water line, the big box volume under water was used as the volume of the water removed. In case of the model without walls under the water line, only the volume of columns and beams under the water line was used. The drag force relates to the relative velocity between tsunami and the object, and is expressed as

$$F_d = \frac{1}{2}\rho_s A C_d U^2 \tag{11.2}$$

where A is the projected area under the water line, C_d the drag force coefficient, and U the relative velocity between tsunami and the object. The entire area below the water line was used as the projected area for the model with walls under the water line, and side-face area of columns below the water line was used as the projected area for the model without walls under the water line. The overall drag force acting on the model with walls becomes about 10 times larger than that of the model without walls.

At the third phase of the analysis, an initial velocity equal to the velocity of tsunami was given to the debris model and the collision to the building was simulated. The drag force initially applied to the debris was zero, because the relative velocity between the debris and tsunami was zero at the initial step. The value reached to about 3.3 MN when the debris perfectly stopped its motion upon impact. For other numerical conditions, the incremental time of the dynamic analysis was set to 1 ms, and Newmark's β method with numerical damping was used as the time-integration scheme.

11.3 NUMERICAL RESULTS

Fig. 11.3 shows the behavior of the building subjected to the Kesennuma wave. These are the moments when the yield function value calculated using Eq. (3.4) had reached to their maximum values. There seemed to be no significant damages due to seismic excitation. Fig. 11.4

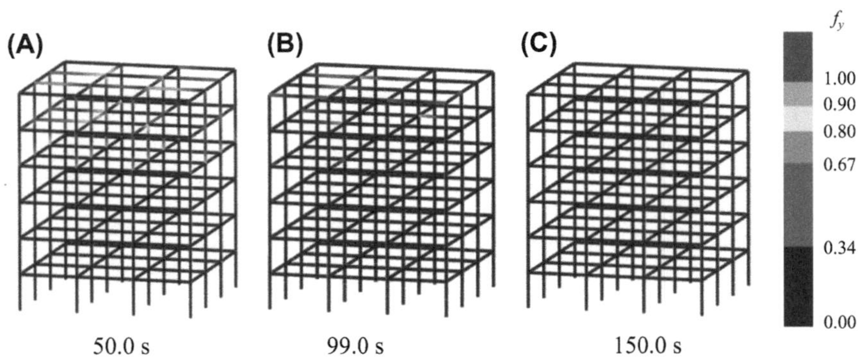

Figure 11.3 Behavior of the building subjected to Kesennuma wave. (A) 50.0 s, (B) 99.0 s, and (C) 150.0 s

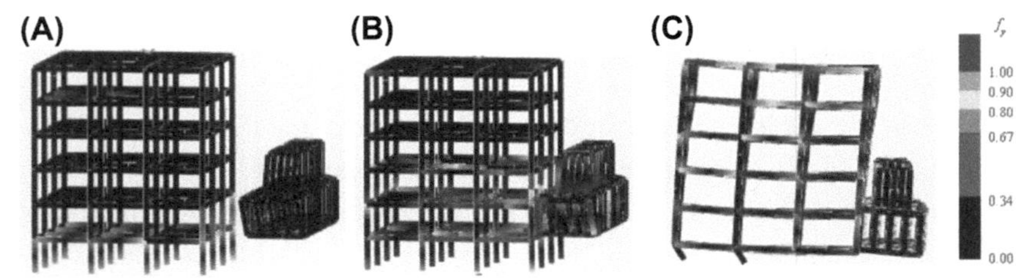

Figure 11.4 Behavior of the building subjected to tsunami and debris collision for the model with walls under the water line. (A) Before impact, (B) after impact, and (C) destruction of base floor.

shows the behavior of the building subjected to tsunami and debris collision for the model with walls under the water line. Because of the walls, the drag force applied is very large and nearly gives a critical damage to the building. The debris collision, however, gives a last trigger to the building to be washed away. After the collision, the building and the debris move with a same velocity and reach to the same value as the tsunami itself. The relative velocity becomes small, and eventually, the drag forces of both decrease to zero as shown in Fig. 11.5.

Fig. 11.6 shows the behavior of the building subjected to tsunami and debris collision for the model without walls under the water line. Since there are no walls under water, the building withstands the fluid force in the first phase. As shown in Fig. 11.7, the drag force applied to the building is smaller than that of the debris in this case. Although a part of the debris separates in the last phase, both models integrate and stop their motion after the impact.

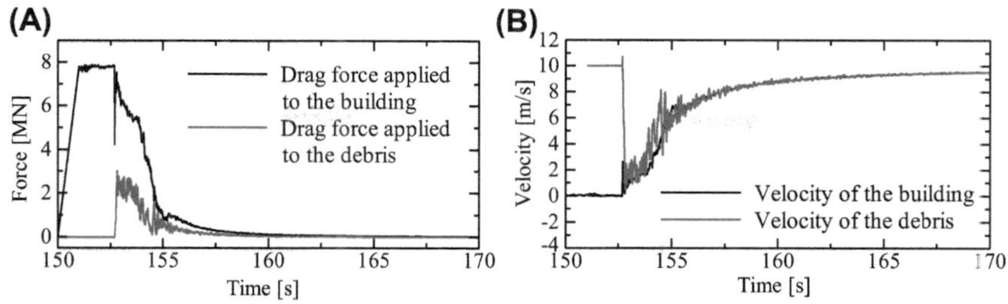

Figure 11.5 Drag forces applied to the building and debris, and their velocities for the model with walls under the water line. (A) Drag forces and (B) velocities.

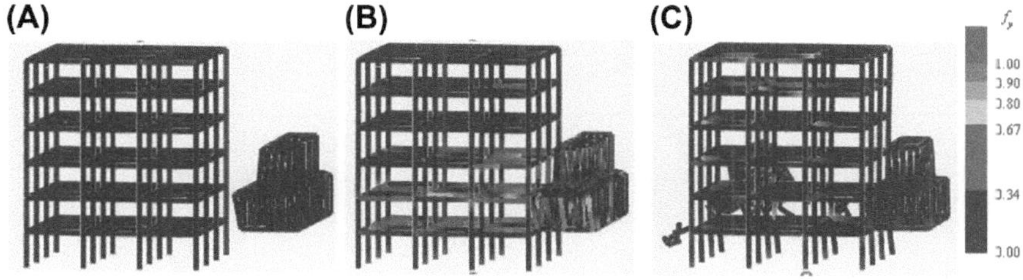

Figure 11.6 Behavior of the building subjected to tsunami and debris collision for the model without walls under the water line. (A) Before impact, (B) after impact, and (C) disintegration of debris.

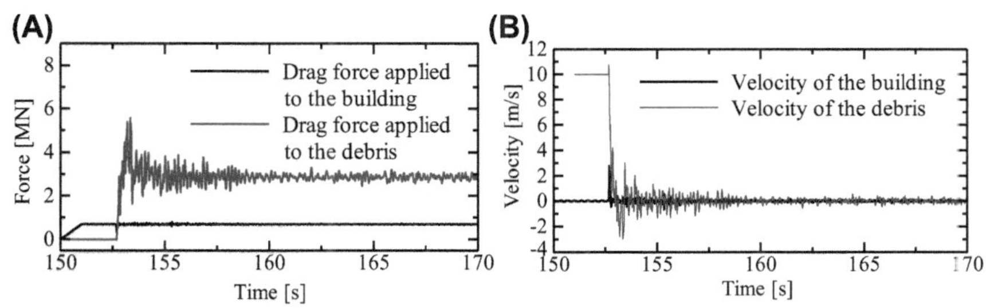

Figure 11.7 Drag forces applied to the building and debris, and their velocities for the model without walls under the water line. (A) Drag forces and (B) velocities.

Furthermore, some head-on collision cases as shown in Fig. 11.8 are carried out; these cases can make the drag force acting on the debris smaller while the impact force on the building may become more intensive. Fig. 11.9 shows the relation between maximum story-drift angle of the building and the velocity of tsunami wave in different inundation heights *d* for the model with walls under the water line. The broken line in the

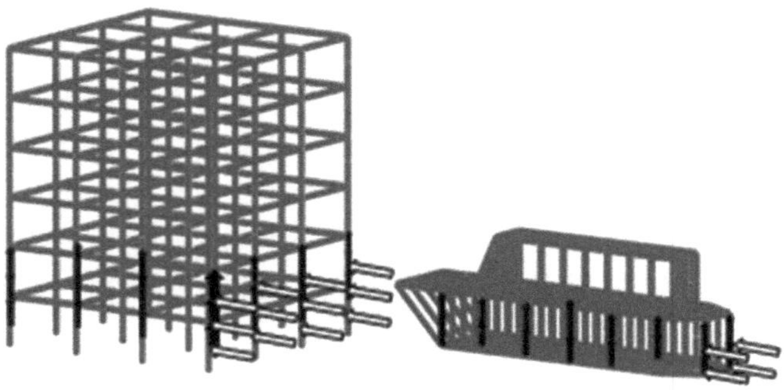

Figure 11.8 Drag forces and buoyant forces applied to the model in head-on collision cases.

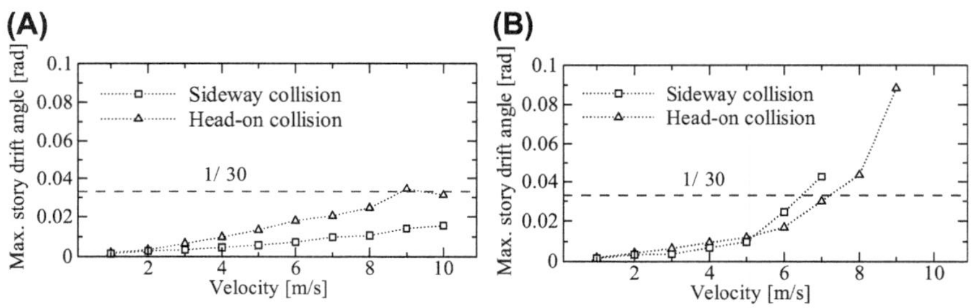

Figure 11.9 Relation between maximum story-drift angle of the building and velocity of tsunami wave in different inundation heights *d* for the model with walls under the water line. (A) $d = 3$ m and (B) $d = 6$ m.

figure indicates the estimated story-drift angle of severe damage, 1/30, which is determined in the structural design guidelines in Japan. If the inundation height is low (Fig. 11.9A), the maximum story-drift angle of the building becomes larger in all cases for the head-on collision. It is because a head-on collision to lower levels of the building gives severe damages to the columns in lower levels and this makes the maximum story-drift angle larger. However, there are some cases in high-inundation heights (Fig. 11.9B), where the damage becomes larger if the debris hits the building sideways at a higher velocity. This is due to the drag force which acts on a broad projected area of the debris in a very high speed flow of tsunami.

As seen in Fig. 11.10, the model without walls under the water line shows similar tendency to that of the former case with walls, although the maximum story-drift angle becomes smaller compared to those in same velocities of the former case. In case of tsunami with low-inundation heights, the damage of the building becomes larger if head-on collision

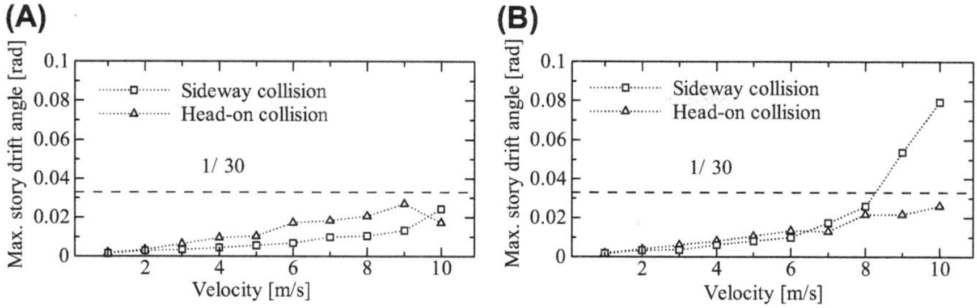

Figure 11.10 Relation between maximum story-drift angle of the building and velocity of tsunami wave in different inundation heights d for the model without walls under the water line. (A) $d = 3$ m and (B) $d = 6$ m.

occurs. In case of a big tsunami with high-inundation height and high velocity, the damage of the building becomes larger if sideway collision occurs.

11.4 SUMMARY

As shown in this chapter, the ASI-Gauss code can easily cope with continuous simulations containing different phases such as seismic excitation, input of fluid forces, and collisions between objects. The numerical results tell us that an appropriate design for tsunami refuge buildings should be estimated and means to avoid the approach of large debris to such buildings should be regarded at the same time.

REFERENCES

Isobe, D., 2017. Simulating collapse behaviors of buildings and motion behaviors of indoor components during earthquakes. In: Zouaghi, T. (Ed.), Earthquakes — Tectonics, Hazard and Risk Mitigation. InTech, ISBN 978-953-51-2885-4, pp. 341–363.

National Institute for Land and Infrastructure Management (NILIM), 2011. Report on Field Surveys and Subsequent Investigations of Building Damage Following the 2011 Off the Pacific Coast of Tohoku Earthquake. Technical Note of National Institute for Land and Infrastructure Management No. 636, Building Research Data No. 132 (in Japanese).

Port and Airport Research Institute (PARI), 2011. Technical Note of the Port and Airport Research Institute. An Investigation Report on the Seismic and Tsunami Damages of Port, Coast and Airport due to 2011 Great East-Japan Earthquake (in Japanese).

CHAPTER TWELVE

Conclusions

12.1 INTRODUCTION

Summaries of the developed numerical codes, applications, and future works are described in this chapter.

12.2 SUMMARY OF THE NUMERICAL CODES

Structural designers usually do not expect a loss of structural members in buildings, but these unexpected events do occur from time to time. The possibilities of simulations considering a loss of structural members are immeasurable, because the risk of lives often depends on the behaviors of structures after they had lost their strengths.

The ASI-Gauss code, described in this book, can practically simulate the collapse behaviors of buildings and motion behaviors of indoor nonstructural components (as shown in the appendices). Of course, the code has limitations. All the structures, including walls and floors, should be modeled with beam elements. One must always be aware that the structures are modeled with extremely approximated elements. However, the computational cost is very low, and the most attractive point of this code is that it can be used in any personal computers with smaller memories.

Although member fracture, contact, and release are not considered, the developed numerical codes can be downloaded from the author's website (Website of ASIFEM; Website of ASIFEM3; Website of ASIRES), as mentioned in other chapters. Structural analysis beginners can start their careers by getting used to the codes (the codes might be revised without notice). The author sincerely wishes their contributions to this field in the near future.

12.3 SUMMARY OF THE APPLICATIONS

Some of the applications described in this book belong to real events occurred in the past. The observed data, if present, are valuable for the validation of the numerical code. One must always realize that validations are

Progressive Collapse Analysis of Structures
ISBN: 978-0-12-812975-3
http://dx.doi.org/10.1016/B978-0-12-812975-3.00012-8

mostly important for these kinds of phenomena with strong nonlinearity, because they are packed with unexpected mechanical interactions. However, it is evident that the numerical code based on a finite element approach can provide useful information such as sectional forces acting in each member and deformation occurring in the structures, while other methods based on non-continuum mechanics cannot. The applicable field of the numerical code is now expanding from collapse analysis of structures to motion analysis of nonstructural components, and it is expected to expand further more.

12.4 FUTURE WORKS

As this practical numerical code for both buildings and indoor components is now available, the author and the members of the research group think that their next goal is to conduct a synthetic analysis which simulates collapse behaviors of buildings and motion behaviors of indoor components, at the same time, by one whole model.

REFERENCES

Website of ASIFEM: Finite element code for framed structures using ASI-Gauss technique. http://www.kz.tsukuba.ac.jp/~isobe/asifem_e.html.

Website of ASIFEM3: Finite element code for framed structures using ASI technique (Cubic beam element version). http://www.kz.tsukuba.ac.jp/~isobe/asifem3_e.html.

Website of ASIRES: Seismic response analysis code for framed structures using ASI-Gauss technique. http://www.kz.tsukuba.ac.jp/~isobe/asires_e.html.

APPENDIX A

Source Program of the ASI-Gauss Code

A.1 INTRODUCTION

The source program of the ASI-Gauss code (ASIFEM) is described in this appendix. As mentioned in the previous chapters, the memory use and computational cost are minimized by subdividing one structural member with only two linear Timoshenko beam elements. Highly accurate solutions and their convergences are guaranteed. Basic loading problems for elastic/elasto-plastic, static/dynamic, infinitesimal/finite deformation, truss/rahmen structures can be analyzed. Total Lagrangian formulation (TLF) and updated Lagrangian formulation (ULF) are both installed as nonlinear incremental theories; however, member fracture, elemental contact, and contact release are not considered. The application files can be downloaded freely from the author's website (Website of ASIFEM).

A.2 USER'S MANUAL FOR THE ASI-GAUSS CODE

A.2.1 Procedure to Make an Input Data

In this section, a general procedure to make an input data is shown. Some of the procedures are similar to those used in many commercial codes; however, one should note that there are some points essential for the ASI-Gauss code.

1. Draw the model you want to analyze, and decide the number of element subdivisions per member. Please note that two elements per member are preferred when using the ASI-Gauss code. In that case, subdivide with two elements between two structural joints in order to obtain highly accurate solutions. Mid-node is to be located at the center of the member. Next, assign the node numbers and element numbers.
2. Fix the coordinates of each node in global Cartesian coordinate system, and write down in input data.

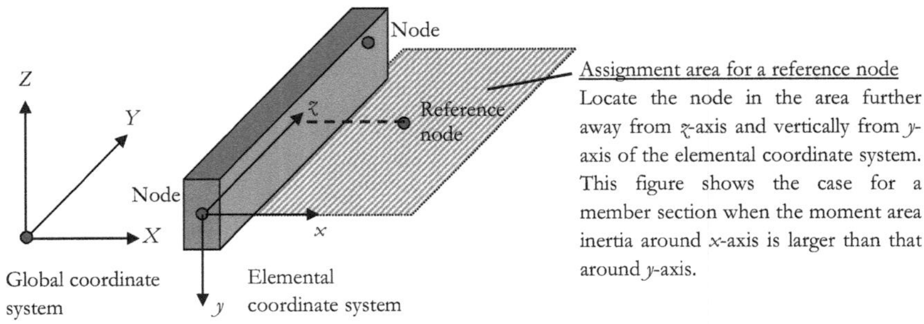

Figure A.1 Coordinates and location of a reference node.

3. Fix the material and sectional properties of structural members, and write down in input data.
4. Fix one reference node per element and write down the coordinates in input data. As shown in Fig. A.1, the reference nodes will set the elemental coordinate system.
5. Write down the connection data of elements (Node No. of two nodes constituting an element, reference node No., material and sectional property No., and pair element No.).
6. Fix the boundary conditions and write down in input data.
7. Write down the total number of elements, nodes, calculation step number and output step number.
8. Fix and write down other numerical conditions for the analysis (e.g., time increment, flags).
9. Set load conditions including consideration of gravity.

After following this procedure, gather up all the input data and save in a file named "`input.txt`" (see Fig. A.2).

A.2.2 Execution Procedure

Place the input data "`input.txt`" in the same folder with "`asifem.exe`", and double click the execution file. When the calculation is successfully executed, three files named as "`output.txt`", "`post.out`" and "`eqs.out`" will be formed. All the input data and outputs will be printed out in "`output.txt`", thus you can check for the input errors by reading this file. "`post.out`" and "`eqs.out`" are the files used for the graphic software "`Graphic.exe`". You can check the outputs graphically by double clicking "`Graphic.exe`". Please read "`manual(Graphic.exe)_e.txt`" for the operation.

<u>Outline of input data (input txt) of FEM code for framed structures (ASIFEM)</u>

16 16 : Total No. of elements, total No. of nodes

1000 100 : No. of calculation steps, No. of output steps

3 2 2 2 1 2 2 1 0 2 : Scheme flag, incremental theory flag, deformation theory flag, elastic/elasto-plastic analysis flag, mass matrix flag, static/dynamic analysis flag, truss/rahmen flag, gravity flag, error message output flag, precision flag (see * for details)
0.44444d+00 0.83333d+00 1.0d-03 : Integration parameters β, δ for Newmark's β method, time increment [s]

1000 1 : No. of steps in loading, No. of nodes in loading
1 1.00d+05 0.0d+00 0.0d+00 0.0d+00 0.0d+00 0.0d+00 : Node No. in loading, subjected forces [N] in X, Y, Z directions of global coordinate system, subjected moments [Nmm] around X, Y, Z-axes

1 : No. of material and sectional properties
2.060d+05 3.000d-01 2.450d+02 2.060d+03 7.900d-06
: Young's modulus [MPa], Poisson's ratio, yield stress [MPa], tangent modulus after yielding [MPa], density [kg/mm^3]

2.500d+03 2.083d+06 1.302d+05 4.391d+05 6.250d+04 1.563d+04
: Cross-sectional area [mm^2], moment of area inertia around x-axis [mm^4], moment of area inertial around y-axis [mm^4], torsional coefficient [mm^4], plastic section modulus for x-direction [mm^3], plastic section modulus for y-direction [mm^3]

24 : No. of fixed degrees of freedom
10 1 10 2 10 3 10 4 10 5 10 6 : (Fixed node No., fixed freedom No.), (Fixed node No., fixed freedom No.), • • •
12 1 12 2 12 3 12 4 12 5 12 6 (Total of 24 pairs of data are aligned in this example)
14 1 14 2 14 3 14 4 14 5 14 6
16 1 16 2 16 3 16 4 16 5 16 6

```
2 1 17 1 2
2 3 18 1 1
4 3 19 1 4
4 5 20 1 3
6 5 21 1 6
6 7 22 1 5
8 7 23 1 8
8 1 24 1 7
9 1 25 1 10
9 10 26 1 9
11 3 27 1 12
11 12 28 1 11
13 5 29 1 14
13 14 30 1 13
15 7 31 1 16
15 16 32 1 15
```

Connection data of elements (specialized for ASI-Gauss technique)
: Node No. (located at the middle of a member), node No. (located at the end of a member), ref. node No., material and sectional property No., pair element No.
(Aligned from the upper row in order as: 1st element, 2nd element, • • • , for 16 elements in this example)

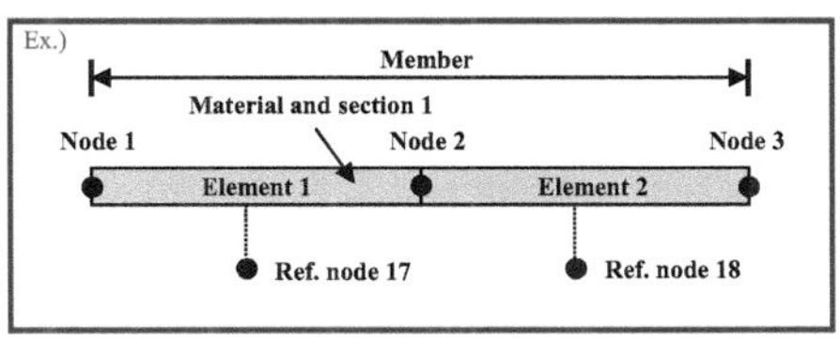


```
1 0.0000d+00 0.0000d+00 2.0000d+03
2 1.0000d+03 0.0000d+00 2.0000d+03
3 2.0000d+03 0.0000d+00 2.0000d+03
4 2.0000d+03 1.0000d+03 2.0000d+03
5 2.0000d+03 2.0000d+03 2.0000d+03
6 1.0000d+03 2.0000d+03 2.0000d+03
7 0.0000d+00 2.0000d+03 2.0000d+03
8 0.0000d+00 1.0000d+03 2.0000d+03
9 0.0000d+00 0.0000d+00 1.0000d+03
10 0.0000d+00 0.0000d+00 0.0000d+00
11 2.0000d+03 0.0000d+00 1.0000d+03
12 2.0000d+03 0.0000d+00 0.0000d+00
13 2.0000d+03 2.0000d+03 1.0000d+03
14 2.0000d+03 2.0000d+03 0.0000d+00
15 0.0000d+00 2.0000d+03 1.0000d+03
16 0.0000d+00 2.0000d+03 0.0000d+00
17 5.0000d+02 -5.0000d+02 2.0000d+03
18 1.5000d+03 -5.0000d+02 2.0000d+03
19 2.5000d+03 5.0000d+02 2.0000d+03
20 2.5000d+03 1.5000d+03 2.0000d+03
21 1.5000d+03 2.5000d+03 2.0000d+03
22 5.0000d+02 2.5000d+03 2.0000d+03
23 -5.0000d+02 1.5000d+03 2.0000d+03
24 -5.0000d+02 5.0000d+02 2.0000d+03
25 0.0000d+00 -5.0000d+02 1.5000d+03
26 0.0000d+00 -5.0000d+02 5.0000d+02
27 2.0000d+03 -5.0000d+02 1.5000d+03
28 2.0000d+03 -5.0000d+02 5.0000d+02
29 2.0000d+03 2.5000d+03 1.5000d+03
30 2.0000d+03 2.5000d+03 5.0000d+02
31 0.0000d+00 2.5000d+03 1.5000d+03
32 0.0000d+00 2.5000d+03 5.0000d+02
```

Coordinates of each node
: Node No., X, Y, Z coordinates in global coordinate system [mm]
(1st ~16th rows are for normal nodes, 17th~32nd rows are for ref. nodes in this example)

Attention: 1.0000d+03 is an expression for double precision of 1.0×10^3

<u>*Details of the flags</u>

Scheme flag	: 1=Conventional method 2=ASI technique 3=ASI-Gauss technique
Incremental theory flag	: 1=TLF 2=ULF
Deformation theory flag	: 1=Infinitesimal deformation 2=Finite deformation
Elastic/elasto-plastic analysis flag	: 1=Elastic analysis 2=Elasto-plastic analysis
Mass matrix flag	: 1=Consistent mass matrix 2=Lumped mass matrix
Static/dynamic analysis flag	: 1=Static analysis 2=Dynamic analysis
Truss/rahmen flag	: 1=Truss structure 2=Rahmen structure
Gravity flag	: 0=No gravity 1=Considering gravity
Error message output flag	: 0=No output 1:Error message output
Precision flag	: 0=High 1:Midium high 2:Medium 3:Low

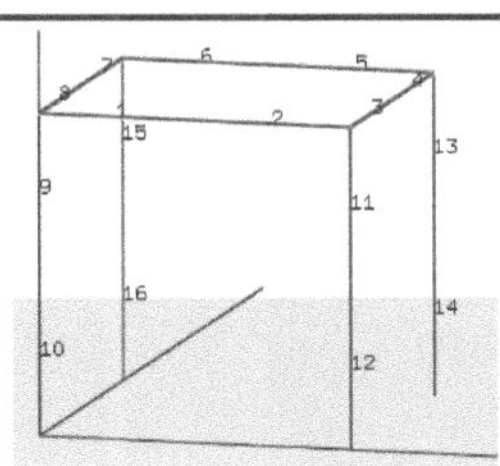

Model of this input data
(No.: element number)

Figure A.2 Outline of the input data for ASIFEM.

A.2.3 How to Handle Errors

There are several ways to cope with execution errors depending on the situation.

1. If your input data is not printed out completely in the file "output.txt", while the data in the printed out area is all right, there might be errors in the next section in the input data.
2. Your input data is completely printed out in the file "output.txt"; however, the results are not printed out. In that case, set the error message output flag on. The file "post.out" for the constructed model will be printed out so that you can view and check the model by using "Graphic.exe".
3. If there seems to be no problem in the model itself viewed by the graphic software, then, read the "Searching for input errors" section in the file "output.txt" which will be printed out when the error message output flag is on. Are all the elements confirmed as "O.K."? If not, check the element information which is not "O.K.". It is highly possible that the coordinates for reference nodes are not properly set.

A.3 SOURCE PROGRAM OF THE ASI-GAUSS CODE

In this section, the main program of the ASI-Gauss code, written in FORTRAN, with some of its main subroutines are shown. The essences of the code are written down as comments in the program. The author has to admit that there might be more efficient ways of programming, so the beginners are encouraged to read the program along with the guidebooks of FORTRAN. The subroutines for the conjugate gradient (CG) method to solve simultaneous equations are omitted. Although it requires a password, the source program can be downloaded from the author's website (Website of ASIFEM).

A.3.1 Main Program

```
!*****************************************************************
      program asifem            !ASI-Gauss code ver.1.3
!*** Main program

!*****************************************************************
```

```
      include 'param.dat'
      dimension coord(mnode+melm,3),dlx(mnode),dly(mnode),
     .    dlz(mnode),el(melm),elori(melm),emate(mmate,mmd),
     .    eqs(melm),eqsout(melm,2),esect(msect,msd),f(mtdf),
     .    fout(mtdf),ind(mnode),ir(melm),
     .    mc(mnc,2),ncr(melm),ne(melm,mdim_ne),
     .    net(mtdf),nte(mtdf),pmx(melm),pmy(melm),pn(melm),
     .    ss(melm),t(melm,3,3),ta(melm,mmtx,mmtx),
     .    tacc(mtdf),te(melm,mdim),tru(melm,3,3),ts(melm,mdim),
     .    tsc(melm,mdim),tsg(melm,mdim),tsl(melm,mdim),
     .    tsr(melm,mdim),tu(mtdf),tuout(mnode,3),tvel(mtdf),
     .    x(maxlena), dltx(mnode),dlty(mnode),dltz(mnode)
!
!*** Open files ***
      open(ii5,status='old',file='input.txt')
      open(ii6,file='post.out')
      open(ii7,file='eqs.out')
      open(ii9,file='output.txt')
!
!*** Initialize parameters and read input data ***
      call inicon(fout,tacc,te,ts,tsc,tsg,tsl,tsr,tu,tuout,tvel,x)
      call input(beta_n,dalfa,dbeta,delta_n,dtime,inielm,nc,ndnod,
     .     nelm,nfree,nlstart,nlstep,nmate,nmetd,nmm,node,
     .     nout,nsd,nsect,nstage,ntdf,coord,dlx,dly,dlz,
     .     emate,esect,ind,ir,mc,ne,pmx,pmy,pn,ss,tuout,
     .     nstr,neg,nerr,neps,dltx,dlty,dltz)
!
!*** Set boundary conditions and initialize parameters for the CG
method ***
      call bdcon(nc,nfree,ntdf,mc,net,nte)
      call cginit(nelm,nfree,node,ntdf,ne,net)
!
!*** Calculation loop (istage=0: A step for calculating self-weight under
gravity) ***
      time=0.0d0
      do istage=0,nstage
!
!*** Make transformation matrix of coordinates ***
        call trans1(istage,nelm,coord,el,elori,ne,t,ta,tru,nerr)
!
!*** Make total mass matrix ***
        call tmasm(nelm,nmm,nfree,ntdf,elori,emate,esect,ne,ta)
!
!*** Make total stiffness matrix ***
        call tstfns(nelm,nfree,ntdf,el,elori,emate,esect,ncr,ne,pmx,
     .          pmy,pn,ss,ta,ts)
!
```

```
!*** Make external force vector and residual force vector ***
        call force(istage,ndnod,nelm,nlstart,nlstep,ntdf,dlx,dly,dlz,
     .           elori,emate,esect,f,ind,ne,nte,neg,dltx,dlty,dltz)
        call resid(nelm,nfree,ntdf,el,f,fout,ne,net,ss,ta,ts,esect)
!
!*** Solve simultaneous equations and obtain displacements, strains and
sectional forces ***
        call solve(beta_n,dalfa,dbeta,delta_n,dtime,istage,nelm,
     .      nfree,nmetd,node,nsd,coord,el,emate,eqs,eqsout,
     .      esect,f,ir,ncr,ne,net,nte,pmx,pmy,pn,ss,t,ta,tacc,
     .      te,tru,ts,tsc,tsg,tsl,tsr,tu,tuout,tvel,x,nstr,neg,
     .      elori,neps)
!
!*** Shift numerical integration points and stress evaluation points ***
        call scont(inielm,nelm,nmetd,ir,ncr,ne,ss,ts,tsc,tsg,tsl,tsr)
!
!*** Print out results ***
        call output(inielm,iout,istage,node,nout,nstage,time,eqsout,
     .      ne,tuout,tu,ts,nelm)
!
!*** Calculate total time ***
        time=time+dtime
      enddo
!
      end
!*****************************************************************
```

*Comments

This program does not contain subroutines for member fracture, elemental contact, and contact release. Important parameters and variables are explained in a subroutine where they first appear. Units for forces are in kgf (input data is written in N) and for lengths are in mm.

A.3.2 Subroutine "bdcon"

```
!*****************************************************************
      subroutine bdcon(nc,nfree,ntdf,mc,net,nte)
!*** Set boundary conditions
!*****************************************************************
      include 'param.dat'
      dimension mc(mnc,2),net(mtdf),nte(mtdf)
!
      do i=1,ntdf
       net(i)=0
       nte(i)=0
      enddo
!
```

```
!*** Make corresponding table ***
!
!*** For fixed DOF ***
      do i=1,nc
       i1=6*(mc(i,1)-1)+mc(i,2)
       nte(nfree+i)=i1
       net(i1)=nfree+i
      enddo
!
!*** For free DOF ***
      k=1
      do i=1,ntdf
       if(net(i)==0) then
        nte(k)=i
        net(i)=k
        k=k+1
       endif
      enddo
!
      return
      end
!*****************************************************************
```

⋆Explanations on variables and arrays

`mc(mnc,2)`: Array of fixed node No. and DOF No. (1 ~ 6)

`net(mtdf)`: Corresponding table from initial DOF No. to DOF No. rearranged by boundary conditions

`nte(mtdf)`: Corresponding table from DOF No. rearranged by boundary conditions to initial DOF No.

A.3.3 Subroutine "bmake"

```
!*****************************************************************
      subroutine bmake(ielm,s,bmatx,el,ta,esect,ne)
!*** Make [B] matrix (generalized strain - nodal displacement matrix)
!*****************************************************************
      include 'param.dat'
      dimension bmatx(mdim,mmtx),el(melm),ta(melm,mmtx,mmtx),
     .    ba(mdim,mmtx),ne(melm,mdim_ne),esect(msect,msd)
!
      do i=1,mdim
        do j=1,mmtx
         ba(i,j)=0.0d0
        enddo
       enddo
!
```

```
!*** [B] matrix for linear Timoshenko beam element (see Eq. (2.8)) ***
      bc=1.0d0/el(ielm)
      ba(1,4)=bc
      ba(1,10)=-bc
      ba(2,5)=-bc
      ba(2,11)=bc
      ba(3,3)=-bc
      ba(3,9)=bc
      ba(4,6)=-bc
      ba(4,12)=bc
      ba(5,1)=-bc
      ba(5,5)=-(1.0d0-s)/2.0d0
      ba(5,7)=bc
      ba(5,11)=-(1.0d0+s)/2.0d0
      ba(6,2)=-bc
      ba(6,4)=(1.0d0-s)/2.0d0
      ba(6,8)=bc
      ba(6,10)=(1.0d0+s)/2.0d0
!
!*** Transformation of coordinates ***
   100 continue
      do i=1,6
       do j=1,12
        sum=0.0d0
        do k=1,12
         sum=sum+ba(i,k)*ta(ielm,k,j)
        enddo
        bmatx(i,j)=sum
       enddo
      enddo
!
      return
      end
!*****************************************************************
```

★Explanations on variables and arrays

ba(mdim,mmtx): Components for [*B*] matrix (see Eq. 2.8)

el(melm): Length of the element at the present incremental step

s: Location of integration point in non-dimensional coordinate

ta(melm,mmtx,mmtx): Transformation matrix from global coordinate system to elemental coordinate system

ne(melm,mdim_ne): Connection data (see Fig. A.2)

esect(msect,msd): Sectional properties of the element (see Fig. A.2)

bmatx(mdim,mmtx): Components for [*B*] matrix after it is transformed (=[*B*]×[*T*])

A.3.4 Subroutine "cdyna"

```
!*******************************************************************
      subroutine cdyna(bt,dalfa,dbeta,delta,dtime,nfree,f,tacc,tvel,
     .     x, neps)
!*** Calculate equation of motion using the CG method
!*******************************************************************
      include 'param.dat'
      dimension f(mtdf),tacc(mtdf),tvel(mtdf),x(maxlena),
     .    w1(maxlena),w2(maxlena),w3(maxlena),w4(maxlena)
!

!*** Incremental equation of motion using Newmark's β method (see
Eq. (2.25)) ***
!  [C]=α[M]+β[K], Rayleigh's damping matrix
      do i=1,nfree
       do j=ilm(i),irma(i)
        gkmat(j)=gkmat(j)+gmmat(j)/(bt*dtime*dtime)
     .      +(dalfa*gmmat(j)+dbeta*gkmat(j))*delta/(bt*dtime)
       enddo'
      enddo
!
      do i=1,nfree
       j=ilm(i)
       f(i)=f(i)+gmmat(j)*tvel(jtab(j))/(bt*dtime)
     .    +gmmat(j)*tacc(jtab(j))*(1.0d0/(2.0d0*bt)-1.0d0)
     .    +(dalfa*gmmat(j)+dbeta*gkmat(j))
     .    *((delta/bt-1.0d0)*tvel(jtab(j))
     .    +(delta/(2.0d0*bt)-1.0d0)*dtime*tacc(jtab(j)))
       do j=ilm(i)+1,irma(i)
        f(i)=f(i)+gmmat(j)*tvel(jtab(j))/(bt*dtime)
     .     +gmmat(j)*tacc(jtab(j))*(1.0d0/(2.0d0*bt)-1.0d0)
     .     +(dalfa*gmmat(j)+dbeta*gkmat(j))
     .     *((delta/bt-1.0d0)*tvel(jtab(j))
     .     +(delta/(2.0d0*bt)-1.0d0)*dtime*tacc(jtab(j)))
        f(jtab(j))=f(jtab(j))+gmmat(j)*tvel(i)/(bt*dtime)
     .         +gmmat(j)*tacc(i)*(1.0d0/(2.0d0*bt)-1.0d0)
     .         +(dalfa*gmmat(j)+dbeta*gkmat(j))
     .         *((delta/bt-1.0d0)*tvel(i)
     .         +(delta/(2.0d0*bt)-1.0d0)*dtime*tacc(i))
       enddo
      enddo
!
!*** Set norm for iteration ***
      if(neps.eq.0) eps=1.0d-14
      if(neps.eq.1) eps=1.0d-6
```

```
      if(neps.eq.2) eps=1.0d-3
      if(neps.eq.3) eps=1.0d-1
      imax=mtdf             ! set iteration limit to a large number
!
!*** CG method ***
      call ddpcg(gkmat,jtab,nfree,ilm,irma,f,x,imax,eps,nstop,
     .     w1,w2,w3,w4)
      if(nstop==2) then
       write(*,'(/1x,36hThe solution is still not converged!)')
       write(ii9,'(/36hThe solution is still not converged!)')
       stop
      endif
!
!*** Acceleration and velocity increments (see Eq. (2.23) and (2.24)) ***
      do i=1,nfree
       dvel=x(i)*delta/(bt*dtime)-tvel(i)*delta/bt
     .      +(1.0d0-delta/(2.0d0*bt))*dtime*tacc(i)
       dacc=x(i)/(bt*dtime*dtime)-tvel(i)/(bt*dtime)-tacc(i)/
     .  (2.0d0*bt)
       tvel(i)=tvel(i)+dvel
       tacc(i)=tacc(i)+dacc
      enddo
!
      return
      end
!**********************************************************************
```

*Explanations on variables and arrays

f(mtdf): Incremental force

tacc(mtdf): Acceleration at each DOF

tvel(mtdf): Velocity at each DOF

x(maxlena): Displacement increment obtained by the CG method

gkmat(maxlenk): Total stiffness matrix rearranged for the CG method

gmmat(maxlenk): Total mass matrix rearranged for the CG method

A.3.5 Subroutine "cdyna_first"

```
!**********************************************************************
      subroutine cdyna_first(nfree,f,x,neps)
!*** Calculate stiffness equation using the CG method

!**********************************************************************
      include 'param.dat'
      dimension f(mtdf),x(maxlena),w1(maxlena),w2(maxlena),
     .     w3(maxlena),w4(maxlena)
!
```

```
!*** Set norm for iteration ***
      if(neps.eq.0) eps=1.0d-14
      if(neps.eq.1) eps=1.0d-6
      if(neps.eq.2) eps=1.0d-3
      if(neps.eq.3) eps=1.0d-1
      imax=mtdf            ! set iteration limit to a large number
!
!*** CG method ***
      call ddpcg(gkmat,jtab,nfree,ilm,irma,f,x,imax,eps,nstop,
     .     w1,w2,w3,w4)
      if(nstop==2) then
       write(*,'(/1x,36hThe solution is still not converged!)')
       write(ii9,'(/36hThe solution is still not converged!)')
       stop
      endif
!
      return
      end
!**********************************************************************
```

★Comments

This subroutine only handles stiffness equation for static analysis.

A.3.6 Subroutine "dmake"

```
!**********************************************************************
      subroutine dmake(am,ielm,dmatx,emate,esect,fd,ncr,ne,pmx,pmy,
     .         pn,q,ts)
!*** Make [D] matrix (generalized sectional force - generalized strain
matrix)
!**********************************************************************
      include 'param.dat'
      dimension dmatx(mdim,mdim),emate(mmate,mmd),esect(msect,msd),
     .     fd(3),ncr(melm),ne(melm,mdim_ne),pmx(melm),pmy(melm),
     .     pn(melm),q(3),ts(melm,mdim)
!
      do i=1,mdim
       do j=1,mdim
        dmatx(i,j)=0.0d0
       enddo
      enddo
!
```

```
!*** [D] matrix for linear Timoshenko beam element (see Eq. (2.12)) ***
      i4=ne(ielm,4)
      G=emate(i4,1)/(2.0d0*(1.0d0+emate(i4,2)))
      dmatx(1,1)=emate(i4,1)*esect(i4,2)    ! EIx Bending stiffness
      around x-axis
      dmatx(2,2)=emate(i4,1)*esect(i4,3)    ! EIy Bending stiffness
      around y-axis
      dmatx(3,3)=emate(i4,1)*esect(i4,1)    ! EA Axial stiffness
      dmatx(4,4)=G*esect(i4,4)              ! GJ Torsional stiffness
      dmatx(5,5)=esect(i4,7)*G*esect(i4,1)  ! αxGA Shearing stiffness
      along x-axis
      dmatx(6,6)=esect(i4,8)*G*esect(i4,1)  ! αyGA Shearing stiffness
      along y-axis
!
!*** Make [D] matrix for plastic deformation (see Eq. (2.20)) ***
      if(ncr(ielm)>0) then
       fd(1)=2.0d0*ts(ielm,1)/pmx(ielm)/pmx(ielm)
       fd(2)=2.0d0*ts(ielm,2)/pmy(ielm)/pmy(ielm)
       fd(3)=2.0d0*ts(ielm,3)/pn(ielm)/pn(ielm)
!
       am=0.0d0
       do i=1,3
        am=am+dmatx(i,i)*fd(i)*fd(i)
       enddo
!
        whard=emate(i4,1)*emate(i4,4)/(emate(i4,1)-emate(i4,4))
        am=am+whard/esect(i4,1)
!
       do i=1,3
        q(i)=dmatx(i,i)*fd(i)
       enddo
       do i=1,3
        do j=1,3
         dmatx(i,j)=dmatx(i,j)-q(i)*q(j)/am
        enddo
       enddo
      endif
!
      return
      end
!*****************************************************************
```

*Explanations on variables and arrays

`dmatx(mdim,mdim)`: Components for [*D*] matrix (see Eq. 2.12)

`emate(mmate,mmd)`: Material properties of the element (see Fig. A.2)

`ncr(melm)`: Elastic/yielding/unloading state flag

0: elastic 1: yielded 2: still yielded -1: unloaded -2: still unloaded

`pmx(melm)`, `pmy(melm)`: Fully plastic bending moments around *x*- and *y*-axes
`pn(melm)`: Fully plastic axial force
`ts(melm,mdim)`: Sectional forces

A.3.7 Subroutine "elmass"

```
!*********************************************************************
      subroutine elmass(ielm,nmm,elori,emate,emmat,esect,ne,ta)
!*** Make elemental mass matrix [ME]

!*********************************************************************
      include 'param.dat'
      dimension elori(melm),emate(mmate,mmd),emmat(mmtx,mmtx),
     .     esect(msect,msd),ne(melm,mdim_ne),ta(melm,mmtx,mmtx),
     .     emt(mmtx,mmtx)
!
      do i=1,mmtx
       do j=1,mmtx
        emmat(i,j)=0.0d0
       enddo
      enddo
!
!*** Make distributed mass matrix for linear Timoshenko beam element (see
Eq. (2.14)) ***
      i4=ne(ielm,4)
      arl=emate(i4,5)*esect(i4,1)*elori(ielm)/(6.0d0*9.8d+03)   !
ρAl/6/9800
      brl=emate(i4,5)*esect(i4,2)*elori(ielm)/(6.0d0*9.8d+03)   !
ρIxl/6/9800
      crl=emate(i4,5)*esect(i4,3)*elori(ielm)/(6.0d0*9.8d+03)   !
ρIyl/6/9800
      drl=brl+crl            ! ρIzl/6/9800
      do i=1,3
       emmat(i,i)=2.0d0*arl
       emmat(i,i+6)=arl
      enddo
      do i=7,9
       emmat(i,i-6)=arl
       emmat(i,i)=2.0d0*arl
      enddo
      emmat(4,4)=2.0d0*brl
      emmat(4,10)=brl
      emmat(10,4)=brl
      emmat(10,10)=2.0d0*brl
      emmat(5,5)=2.0d0*crl
      emmat(5,11)=crl
      emmat(11,5)=crl
```

```
      emmat(11,11)=2.0d0*crl
      emmat(6,6)=2.0d0*drl
      emmat(6,12)=drl
      emmat(12,6)=drl
      emmat(12,12)=2.0d0*drl
!
!*** Transformation of coordinates ***
      do i=1,12
       do j=1,12
        sum=0.0d0
        do k=1,12
         sum=sum+emmat(i,k)*ta(ielm,k,j)
        enddo
        emt(i,j)=sum
       enddo
      enddo
      do i=1,12
       do j=1,12
        sum=0.0d0
        do k=1,12
         sum=sum+ta(ielm,k,i)*emt(k,j)
        enddo
        emmat(i,j)=sum        ! emmat=[T]T[M][T]
       enddo
      enddo
!
      return
      end
!*****************************************************************
```

*Explanations on variables and arrays

elori(melm): Initial element length

emmat(mmtx,mmtx): Components for distributed mass matrix (see Eq. 2.14)

emt(mmtx,mmtx): Components for $[M_E]$ matrix after it is transformed to global coordinate system

A.3.8 Subroutine "elstif"

```
!*****************************************************************

      subroutine elstif(ielm,ekmat,el,elori,emate,esect,ncr,ne,pmx,
     .       pmy,pn,ss,ta,ts)
!*** Make elemental stiffness matrix [KE]

!*****************************************************************
```

```
      include 'param.dat'
      dimension ekmat(mmtx,mmtx),el(melm),elori(melm),
     .     emate(mmate,mmd),esect(msect,msd),ncr(melm),
     .     pmx(melm),pmy(melm),pn(melm),ss(melm),q(3),
     .     ta(melm,mmtx,mmtx),ts(melm,mdim),fd(3),
     .     bmatx(mdim,mmtx),db(mdim,mmtx),dmatx(mdim,mdim),
     .     ekb(mmtx,mmtx),ne(melm,mdim_ne)
!
      do i=1,mmtx
       do j=1,mmtx
        ekb(i,j)=0.0d0
       enddo
      enddo
      s=ss(ielm)
      call bmake(ielm,s,bmatx,el,ta,esect,ne)
      call dmake(am,ielm,dmatx,emate,esect,fd,ncr,ne,pmx,pmy,pn,q,ts)
!
!*** Make elemental stiffness matrix [KE]=[B]T[D][B] ***
      do i=1,6
       do j=1,12
        sum=0.0d0
        do k=1,6
         sum=sum+dmatx(i,k)*bmatx(k,j)
        enddo
        db(i,j)=sum
       enddo
      enddo
      do i=1,12
       do j=1,12
        sum=0.0d0
        do k=1,6
         sum=sum+bmatx(k,i)*db(k,j)
        enddo
        ekb(i,j)=sum
       enddo
      enddo
!
      do i=1,12
       do j=1,12
        ekmat(i,j)=ekb(i,j)*el(ielm)
       enddo
      enddo
!
      return
      end
!*****************************************************************
```

*Explanations on variables and arrays

`ekmat(mmtx,mmtx)`: Components for elemental stiffness matrix

`ss(melm)`: Location of numerical integration point in each element

`db(mdim,mmtx)`: $=[D][B]$

`ekb(mmtx,mmtx)`: $=[B]^{\mathrm{T}}[D][B]$

A.3.9 Subroutine "`force`"

```
!*****************************************************************
      subroutine force(istage,ndnod,nelm,nlstart,nlstep,ntdf,dlx,
     .         dly,dlz,elori,emate,esect,f,ind,ne,nte,neg,
     .         dltx,dlty,dltz)
!*** Set external forces
!*****************************************************************
      include 'param.dat'
      dimension dlx(mnode),dly(mnode),dlz(mnode),elori(melm),
     .     emate(mmate,mmd),esect(msect,msd),f(mtdf),ftot(mtdf),
     .     ind(mnode),ne(melm,mdim_ne),nte(mtdf),
     .     dltx(mnode),dlty(mnode),dltz(mnode)
!
      emass=0.0d0
      do i=1,mtdf
       ftot(i)=0.0d0
      enddo
!
!*** Calculate self-weight under gravity at istage=0 ***
      if(istage==0) then
       do ielm=1,nelm
        i4=ne(ielm,4)
        if(neg==1) emass=-emate(i4,5)*esect(i4,1)*elori(ielm)/2.0d0!
     .    emass=-ρAl/2
        ftot(6*ne(ielm,1)-3)=ftot(6*ne(ielm,1)-3)+emass
        ftot(6*ne(ielm,2)-3)=ftot(6*ne(ielm,2)-3)+emass
       enddo
      endif
!
!*** Calculate incremental external forces ***
      if(nlstart<=istage.and.istage<nlstart+nlstep) then
       do i=1,ndnod
        ftot(6*ind(i)-5)=dlx(ind(i))/dble(nlstep)
        ftot(6*ind(i)-4)=dly(ind(i))/dble(nlstep)
        ftot(6*ind(i)-3)=dlz(ind(i))/dble(nlstep)
        ftot(6*ind(i)-2)=dltx(ind(i))/dble(nlstep)
        ftot(6*ind(i)-1)=dlty(ind(i))/dble(nlstep)
        ftot(6*ind(i))=dltz(ind(i))/dble(nlstep)
       enddo
      endif
!
```

```
!*** Rearrange DOF No. ***
      do i=1,ntdf
       i1=nte(i)
       f(i)=ftot(i1)
      enddo
!
      return
      end
!*****************************************************************
```

*Explanations on variables and arrays

`dlx(mnode),dly(mnode),dlz(mnode)`: Total load applied to the node in *X*, *Y* and *Z* directions

`dltx(mnode),dlty(mnode),dltz(mnode)`: Total moment applied to the node around *X*, *Y* and *Z* axes

`ftot(mtdf)`: Temporary array for incremental external forces

`emass`: Half of the element mass acted on vertically downward direction

A.3.10 Subroutine "inicon"

```
!*****************************************************************
      subroutine inicon(fout,tacc,te,ts,tsc,tsg,tsl,tsr,tu,tuout,
     .     tvel,x)
!*** Initialize each parameter

!*****************************************************************
      include 'param.dat'
      dimension fout(mtdf),tacc(mtdf),te(melm,mdim),tvel(mtdf),
     .   ts(melm,mdim),tsc(melm,mdim),tsg(melm,mdim),x(maxlena),
     .   tsl(melm,mdim),tsr(melm,mdim),tu(mtdf),tuout(mnode,3)
!
!*** Initialize variables and arrays ***
      do i=1,mtdf
       fout(i)=0.0d0
       tu(i)=0.0d0
       tvel(i)=0.0d0
       tacc(i)=0.0d0
      enddo
      do i=1,mnode
       do j=1,3
        tuout(i,j)=0.0d0
       enddo
      enddo
      do i=1,maxlena
       x(i)=0.0d0
      enddo
```

```
      do i=1,melm
       do j=1,mdim
        te(i,j)=0.0d0
        ts(i,j)=0.0d0
        tsc(i,j)=0.0d0
        tsg(i,j)=0.0d0
        tsl(i,j)=0.0d0
        tsr(i,j)=0.0d0
       enddo
      enddo
!
      return
      end
!*****************************************************************
```

*Comments

These variables and arrays should be initialized in order to calculate incrementally.

A.3.11 Subroutine "input"

```
!*****************************************************************
      subroutine input(beta_n,dalfa,dbeta,delta_n,dtime,inielm,nc,
     .       ndnod,nelm,nfree,nlstart,nlstep,nmate,nmetd,
     .       nmm,node,nout,nsd,nsect,nstage,ntdf,coord,
     .       dlx,dly,dlz,emate,esect,ind,ir,mc,ne,pmx,pmy,
     .       pn,ss,tuout,nstr,neg,nerr,neps,dltx,dlty,dltz)
!*** Read input data

!*****************************************************************
      include 'param.dat'
      dimension coord(mnode+melm,3),dlx(mnode),dly(mnode),dlz(mnode),
     .     emate(mmate,mmd),esect(msect,msd),ind(mnode),ir(melm),
     .     mc(mnc,2),ne(melm,mdim_ne),pmx(melm),pmy(melm),pn(melm),
     .     ss(melm),tuout(mnode,3),temp(melm,11),dltx(mnode),
     .     dlty(mnode),dltz(mnode)
!
      rewind ii5
!
!*** Read basic information ***
      read(ii5,*) nelm,node,nstage,nout
      read(ii5,*) nmetd,nut,nif,nep,nmm,nsd,nstr,neg,nerr,neps
      read(ii5,*) beta_n,delta_n,dtime
      dalfa=0.0d0    ! Rayleigh's damping initially not implemented
      dbeta=0.0d0    ! Rayleigh's damping initially not implemented
      inielm=nelm    ! Save initial total No. of elements
      ntdf=6*node    ! Calculate total No. of DOF
      if(nstage/=0.and.nout>nstage) nout=nstage
      if(nstage==0.or.nout==0) nout=1
!
```

```
!*** Print out basic information ***
      write(ii9,900)
  900 format('<< STRUCTURAL ANALYSIS OF FRAMED STRUCTURE >>')
      write(ii9,901)
  901 format(/,'<< ANALYTICAL CONDITIONS >>')
      write(ii9,902) nep
  902 format(/,'Elasto-plasticity =',i6,5x,'(1:Elastic,
     .    2:Elasto-plastic)')
      write(ii9,903) nif
  903 format('Deformation theory =',i5,5x,
     .    '(1:Infinitesimal,2:Finite)')
      write(ii9,904) nut
  904 format('Incremental theory =',i5,5x,'(1:T.L.F., 2:U.L.F.)')
      write(ii9,905) nsd
  905 format('Static or Dynamic =',i6,5x,'(1:Static, 2:Dynamic)')
       write(ii9,906) nstr
  906 format('Structure =',i14,5x,'(1:Truss, 2:Rahmen)')
       write(ii9,907) nmetd
  907 format('Method =',i17,5x,
     .        '(1:Conventional scheme, 2:ASI, 3:ASI-Gauss)')
      write(ii9,942) nerr
  942 format('Output =',i17,5x,
     .    '(0:Normal mode, 1:Error message output mode)')
      write(ii9,943) neps
  943 format('Precision =',i14,5x,
     .    '(0:High, 1:Medium high, 2:Medium, 3:Low)')
!
      if(nsd==2) then
       write(ii9,908)
  908 format(/,'<< DYNAMIC PARAMETERS >>')
       write(ii9,909) nmm
  909 format(/,'Mass matrix =',i12,5x,'(1:Consistent, 2:Lumped)')
       write(ii9,910) beta_n,delta_n,dtime
  910 format('Newmark β =',f15.3,/,'Newmark δ =',f15.3,
     .    /,'Time increments [s] =',d12.5)
      endif
!
      write(ii9,911)
  911 format(/,'<< OUTPUT CONDITIONS >>')
      write(ii9,912) nstage,nout
  912 format(/,'No. of steps =',i11,/,'No. of outputs =',i9)
!
!*** Initialize locations of numerical integration points and stress
evaluation points ***
      do ielm=1,nelm
       if(nmetd==3) then     ! ASI-Gauss technique
        ss(ielm)=1.0d0-2.0d0/dsqrt(3.0d0)
```

```
        else                    ! Conventional FEM and ASI technique
         ss(ielm)=0.0d0
        endif
        ir(ielm)=0
       enddo
!
!*** Read external forces ***
       read(ii5,*) nlstep,ndnod
       nlstart=1
       do i=1,ndnod
        read(ii5,*) ind(i),dlx(ind(i)),dly(ind(i)),dlz(ind(i)),
     .       dltx(ind(i)),dlty(ind(i)),dltz(ind(i))
        dlx(ind(i))=dlx(ind(i))/9.8d0
        dly(ind(i))=dly(ind(i))/9.8d0
        dlz(ind(i))=dlz(ind(i))/9.8d0
        dltx(ind(i))=dltx(ind(i))/9.8d0
        dlty(ind(i))=dlty(ind(i))/9.8d0
        dltz(ind(i))=dltz(ind(i))/9.8d0
       enddo
!
!*** Print out external forces ***
       write(ii9,936)
   936 format(/,'<< LOAD CONDITIONS >>')
       write(ii9,937) nlstart,nlstep,ndnod
   937 format(/,'Initial step to start loading =',i5,
     .    /,'Loading steps =',i21,/,'No. of nodes in loading =',i11)
       write(ii9,938) neg
   938 format('Dead load due to gravity =',i10,5x,
     .    '(0:No gravity, 1:Considering gravity)')
       write(ii9,939)
   939 format(/,'Node No.',3x,'Force(X) [N]',3x,'Force(Y) [N]',3x,
     .     'Force(Z) [N]',2x,'Moment(X)[Nmm]',1x,'Moment(Y)[Nmm]',1x,
     .     'Moment(Z)[Nmm]')
       do i=1,ndnod
       write(ii9,940) ind(i),dlx(ind(i))*9.8d0,dly(ind(i))*9.8d0,
     .       dlz(ind(i))*9.8d0,dltx(ind(i))*9.8d0,
     .       dlty(ind(i))*9.8d0,dltz(ind(i))*9.8d0
       enddo
   940 format(i4,3x,6d15.5)
!
!*** Read material and sectional properties ***
       read(ii5,*) nmate
       nsect=nmate
       read(ii5,*) ((temp(i,j),j=1,11),i=1,nmate)
       do i=1,nmate
        emate(i,1)=temp(i,1)/9.8d0
        emate(i,2)=temp(i,2)
```

```
         emate(i,3)=temp(i,3)/9.8d0
         emate(i,4)=temp(i,4)/9.8d0
         emate(i,5)=temp(i,5)
        enddo
        do i=1,nsect
         esect(i,1)=temp(i,6)
         esect(i,2)=temp(i,7)
         esect(i,3)=temp(i,8)
         esect(i,4)=temp(i,9)
         esect(i,5)=temp(i,10)
         esect(i,6)=temp(i,11)
         esect(i,7)=5.0d0/6.0d0
         esect(i,8)=5.0d0/6.0d0
        enddo
!
!*** Print out material and sectional properties ***
        write(ii9,913)
  913 format(/,'<< MEMBER CONDITIONS >>')
        write(ii9,914)
  914 format(/,'-- Material properties --')
        write(ii9,915) nmate
  915 format('No. of materials =',i7)
        write(ii9,916)
  916 format('Material No.',3x,'E [MPa]',10x,'ν',
     .    10x,'σy [MPa]',7x,'Et [MPa]',4x,'ρ [kg/mm^3]')
        do i=1,nmate
         write(ii9,917) i,emate(i,1)*9.8d0,emate(i,2),emate(i,3)*9.8d0,
     .         emate(i,4)*9.8d0,emate(i,5)
  917 format(i8,5d15.5)
        enddo
        write(ii9,918)
  918 format(/,'-- Sectional properties --')
        write(ii9,919) nsect
  919 format('No. of sections =',i8)
        write(ii9,920)
  920 format('Section No.',4x,'A [mm^2]',6x,'Ix [mm^4]',
     . 6x,'Iy [mm^4]',6x,'J [mm^4]',6x,'Zpx [mm^3]',5x,'Zpy [mm^3]')
        do i=1,nsect
         write(ii9,921) i,(esect(i,j),j=1,6)
  921 format(i8,6d15.5)
        enddo
!
!*** Read fixed conditions ***
        read(ii5,*) nc
        read(ii5,*) ((mc(i,j),j=1,2),i=1,nc)
        nfree=ntdf-nc
!
```

```
!*** Print out fixed conditions ***
      write(ii9,922)
  922 format(/,'<< MODEL CONDITIONS >>')
      write(ii9,923) nelm,node
  923 format(/,'No. of elements =',i8,/,'No. of nodes =',i11)
      write(ii9,932)
  932 format(/,'-- Fixed degrees of freedom --')
      write(ii9,933) nc
  933 format('No. of fixed degrees =',i5)
      write(ii9,934)
  934 format('Fixed degree No.',2x,'Node No.',2x,'Degree No.')
      do i=1,nc
      write(ii9,935) i,(mc(i,j),j=1,2)
  935 format(i4,8x,2i10)
      enddo
!
!*** Read connection data ***
      read(ii5,*) ((temp(ielm,j),j=1,5),ielm=1,nelm)
      do ielm=1,nelm
       do j=1,4
        ne(ielm,j)=temp(ielm,j)
       enddo
       ne(ielm,5)=1
       ne(ielm,6)=temp(ielm,5)
      enddo
!
!*** Print out connection data ***
      write(ii9,929)
  929 format(/,'-- Node connections --')
      write(ii9,930)
  930 format('Element No.',7x,'Connected nodes',6x,
     .'Ref.node',3x,'Mat.Sect.No.',2x,'Pair element')
      do i=1,nelm
       write(ii9,931) i,(ne(i,j),j=1,4),ne(i,6)
  931 format(i4,4x,5i12)
      enddo
!
!*** Calculate fully plastic sectional forces ***
      do ielm=1,nelm
       i4=ne(ielm,4)
       G=emate(i4,1)/(2.0d0*(1.0d0+emate(i4,2)))
       pmx(ielm)=emate(i4,3)*esect(i4,5) ! Fully plastic moment around
       x axis (=σyZpx)
       pmy(ielm)=emate(i4,3)*esect(i4,6) ! Fully plastic moment around
       y axis (=σyZpy)
       pn(ielm)=emate(i4,3)*esect(i4,1) ! Fully plastic axial force
       (=σyA)
```

```
        if(nep==1) then    ! Large values are applied to avoid yielding if
                           ! elastic
         pmx(ielm)=pmx(ielm)*1.0d+10
         pmy(ielm)=pmy(ielm)*1.0d+10
         pn(ielm)=pn(ielm)*1.0d+10
        endif
       enddo
!
!*** Read initial coordinates of nodes and reference nodes ***
       do inode=1,node+nelm
        read(ii5,*) i,(coord(i,j),j=1,3)
       enddo
       do i=1,node
        do j=1,3
         tuout(i,j)=coord(i,j)
        enddo
       enddo
!
!*** Print out initial coordinates of nodes and reference nodes ***
       write(ii9,924)
   924 format(/,'-- Node coordinates --')
       write(ii9,925)
   925 format('Node No.',7x,'X [mm]',9x,'Y [mm]',9x,'Z [mm]')
       do i=1,node
        write(ii9,926) i,(coord(i,j),j=1,3)
   926 format(i4,3x,3d15.5)
       enddo
       write(ii9,927)
   927 format('-------------------------------- Reference nodes')
       do i=node+1,node+nelm
        write(ii9,926) i,(coord(i,j),j=1,3)
       enddo
       write(ii9,928)
   928 format('----------------------------------------')
       write(ii9,941)
   941 format(/,'*********************************************
      .**************************************')
!
!*** Print out graphics data for checking the constructed model ***
       if(nerr==1) then
        write(ii6,'(2i8,x,f12.5)') inielm,node,-1.0
        write(ii6,'(12i7)') ((ne(ielm,j),j=1,2),ielm=1,inielm)
       do inode=1,node
        write(ii6,'(i7,6e13.5)') inode,(coord(inode,j),j=1,3)
       enddo
```

```
      do ielm=1,inielm
       write(ii7,*) 0.0,0.0
      enddo
      endif
!
      return
      end
!*********************************************************************
```

*Explanations on variables and arrays

`nelm`: Total No. of elements

`node`: Total No. of nodes

`nstage`: Total No. of steps

`nout`: Total No. of output steps

`nmetd`: 1: conventional FEM, 2: ASI technique, 3: ASI-Gauss technique

`nut`: 1: TLF 2: ULF (fixed to nut=2 in this appendix)

`nif`: 1: infinitesimal deformation 2: finite deformation (fixed to nif=1 in this appendix)

`nep`: 1: elastic analysis, 2: elasto-plastic analysis

`nsd`: 1: static analysis, 2: dynamic analysis

`nstr`: 1: truss structure, 2: rahmen structure

`neg`: 0: no gravity, 1: under gravity

`nerr`: 0: no error messages, 1: print out error messages

`neps`: 0: high accuracy, 1: middle-high, 2: middle, 3: low

`beta_n`: Newmark's β

`delta_n`: Newmark's δ

`dtime`: Time increment

`nlstart`: Step No. to start loading

`nlstep`: Total load step No.

`ndnod`: Total node No. subjected to loading

`ind(mnode)`: Node No. subjected to loading

`nmate`: Total variety No. of materials

`emate(i,1)`: Young's modulus E of No. i material [N/mm^2](=[MPa])

`emate(i,2)`: Poisson's ratio ν of No. i material

`emate(i,3)`: Yield stress σ_y of No. i material [N/ mm^2](=[MPa])

`emate(i,4)`: Tangent modulus of plasticity E_t of No. i material [N/mm^2](=[MPa])

`emate(i,5)`: Density ρ of No. i material [kg/mm^3]

`nsect`: Total variety No. of sections

`esect(i,1)`: Cross-sectional area A [mm^2]

`esect(i,2)`: Moment of area inertia around x-axis I_x [mm^4]

esect(i,3): Moment of area inertia around y-axis I_y [mm^4]
esect(i,4): Torsinoal coefficient J [mm^4]
esect(i,5): Plastic section modulus in x direction Z_{px} [mm^3]
esect(i,6): Plastic section modulus in y direction Z_{py} [mm^3]
esect(i,7): Shear correction factor in x direction α_x
esect(i,8): Shear correction factor in y direction α_y
nc: Total No. of fixed DOF
mc(i,1): Fixed node No.
mc(i,2): Fixed freedom No. (1 ~ 6)
ntdf: Total No. of DOF
nfree: Total No. of free nodes
ne(i,1),ne(i,2): Connection data of the element (see Fig. A.2)
ne(i,3): Reference node No. of the element (see Fig. A.2)
ne(i,4): Material property No. and sectional property No. of the element
ne(i,5): Category of the element (=1 in this program)
ne(i,6): Element No. of the pair element constituting the same member
coord(i,1),coord(i,2),coord(i,3): Initial X, Y, Z coordinates of node No. i
ir(melm): Location of stress evaluation point in each element
tuout(mnode,3): Coordinates of the node at the present step

A.3.12 Subroutine "output"

```
!*****************************************************************
      subroutine output(inielm,iout,istage,node,nout,nstage,time,
     .         eqsout,ne,tuout,tu,ts,nelm)
!*** Print out calculation results
!*****************************************************************
      include 'param.dat'
      dimension eqsout(melm,2),ne(melm,mdim_ne),tuout(mnode,3),
     .      tu(mtdf),ts(melm,mdim)
!
      if(istage==0) iout=0
      if(nstage/=0.and.mod(istage,nstage/nout)/=0) return
!
!*** Print out information required for post processing ***
      iout=iout+1
      write(ii6,'(2i8,x,f12.5)') inielm,node,time
      write(ii6,'(12i7)') ((ne(ielm,j),j=1,2),ielm=1,inielm)
      do inode=1,node
```

```
         write(ii6,'(i7,6e13.5)') inode,(tuout(inode,j),j=1,3)
        enddo
        do ielm=1,inielm
         write(ii7,*) (eqsout(ielm,j),j=1,2)
        enddo
        write(*,'(i7,3x,f10.3,3x,i5)') istage,time,iout
!
!*** Print out step No. and total time ***
        if(istage==0) write(ii9,941)
    941 format(/,'**********************************************
       .**********************************')
        if(istage==0) write(ii9,909)
        write(ii9,910) istage,time
    909 format('<< OUTPUTS >>')
    910 format(/2x,8histage =,i10,3x,8h time =,e15.7,3x)
!
!*** Print out displacements ***
        write(ii9,*) '-- Displacements --'
        write(ii9,919)
        do 200 i=1,node
         write(ii9,920) i,tu(6*i-5),tu(6*i-4),tu(6*i-3)
    200 continue
    919 format('Node No.',5x,'X [mm]',10x,'Y [mm]',10x,'Z [mm]')
    920 format(i4,3d16.4)
!
!*** Print out sectional forces ***
        write(ii9,*) '-- Sectional forces --'
        write(ii9,929)
        do 300 i=1,nelm
         write(ii9,930) i,(ts(i,j)*9.8d0,j=1,6)
    300 continue
    929 format('Element No.',1x,'Mx [Nmm]',8x,'My [Nmm]',9x,
       .'N [N]',10x,'Mt [Nmm]',9x,'Vx [N]',10x,'Vy [N]')
    930 format(i4,6d16.4)
!
        return
        end
!***********************************************************
```

⋆Explanations on variables and arrays

`eqsout(melm,2)`: Yield function values required for post processing

`tu(mtdf)`: Displacements at each DOF

A.3.13 Subroutine "resid"

```
!*****************************************************************
      subroutine resid(nelm,nfree,ntdf,el,f,fout,ne,net,ss,
     .   ta,ts,esect)
!*** Calculate external, internal and residual forces
!*****************************************************************
      include 'param.dat'
      dimension el(melm),f(mtdf),fout(mtdf),ne(melm,mdim_ne),
     .     net(mtdf),ss(melm),ta(melm,mmtx,mmtx),ts(melm,mdim),
     .     bmatx(mdim,mmtx),btdg(mmtx),fint(mtdf),fres(mtdf),
     .     esect(msect,msd)
!
      do i=1,ntdf
       fint(i)=0.0d0
       fres(i)=0.0d0
      enddo
!
!*** Calculate internal forces ***
      do ielm=1,nelm
       s=ss(ielm)
      call bmake(ielm,s,bmatx,el,ta,esect,ne)
       do j=1,12
        sum=0.0d0
        do k=1,6
         sum=sum+bmatx(k,j)*ts(ielm,k)
        enddo
        btdg(j)=sum*el(ielm)
       enddo
       kevab=0
       do i=1,2
        nposn=6*(ne(ielm,i)-1)
        do idofn=1,6
         kevab=kevab+1
         nposn=nposn+1
         l1=net(nposn)
         if(l1>nfree) cycle
         fint(l1)=fint(l1)+btdg(kevab)
        enddo
       enddo
      enddo
!
```

```
!*** Calculate external, residual and incremental forces ***
      do i=1,nfree
       fres(i)=fout(i)-fint(i)
       fout(i)=fout(i)+f(i)
       f(i)=f(i)+fres(i)
      enddo
!
      return
      end
!*****************************************************************
```

★Explanations on variables and arrays

`fout(mtdf)`: External forces applied at each DOF

`btdg(mmtx)`: $\int [B]^{\mathrm{T}}[\sigma]\mathrm{d}l$

`fint(mtdf)`: Internal forces obtained at each DOF

`fres(mtdf)`: Residual forces obtained at each DOF

A.3.14 Subroutine "`scont`"

```
!*****************************************************************
      subroutine scont(inielm,nelm,nmetd,ir,ncr,ne,ss,ts,
     .   tsc,tsg,tsl,tsr)
!*** Control numerical integration points and stress evaluation points
!*****************************************************************
      include 'param.dat'
      dimension ir(melm),ncr(melm),ne(melm,mdim_ne),ss(melm),
     .     ts(melm,mdim),tsc(melm,mdim),tsg(melm,mdim),
     .     tsl(melm,mdim),tsr(melm,mdim)
!
!*** Shift numerical integration points and stress evaluation points ***
      if(nmetd==1) then    ! Conventional FEM
       do ielm=1,nelm
        ss(ielm)=0.0d0
        do j=1,6
         ts(ielm,j)=tsc(ielm,j)
        enddo
       enddo
       return
      else                         ! ASI and ASI-Gauss techniques
       do ielm=1,inielm
        if(ncr(ielm)>0) then    ! For elements in plastic range
         ss(ielm)=dble(-ir(ielm))
```

```
          do j=1,6
           if(ir(ielm)==-1) ts(ielm,j)=tsl(ielm,j)
           if(ir(ielm)==1) ts(ielm,j)=tsr(ielm,j)
          enddo
          cycle
         elseif(ncr(ielm)<=0) then        ! For elements in elastic rarge
          ss(ielm)=0.0d0
          do j=1,6
           ts(ielm,j)=tsc(ielm,j)
          enddo
         endif
         if(nmetd==2) cycle               ! For ASI-Gauss technique only
          i6=ne(ielm,6)
          if(ncr(ielm)<=0.and.ncr(i6)<=0) then
                                          ! If pair elements are both in
          elastic range
          ss(ielm)=1.0d0-2.0d0/dsqrt(3.0d0) ! See Fig. 3.1
          do j=1,6
           ts(ielm,j)=tsg(ielm,j)
          enddo
          endif
        enddo
        do ielm=inielm+1,nelm
         ss(ielm)=0.0d0
         do j=1,6
          ts(ielm,j)=tsc(ielm,j)
         enddo
        enddo
       endif
!
       return
       end
!*******************************************************************
```

*Explanations on variables and arrays

`tsc(melm,mdim)`: Sectional forces calculated at the mid-point (s=0) of the element

`tsg(melm,mdim)`: Sectional forces calculated at the Gaussian integration points ($s=\pm 1/\sqrt{3}$) of the member

`tsl(melm,mdim)`: Sectional forces calculated at the starting node (s=-1) of the element or mid-point of the member (see Fig. 3.1)

`tsr(melm,mdim)`: Sectional forces calculated at the ending node (s=1) of the element or either end of the member (see Fig. 3.1)

A.3.15 Subroutine "solve"

```
!*****************************************************************
      subroutine solve(beta_n,dalfa,dbeta,delta_n,dtime,istage,
     .       nelm,nfree,nmetd,node,nsd,coord,el,emate,
     .       eqs,eqsout,esect,f,ir,ncr,ne,net,nte,pmx,pmy,
     .       pn,ss,t,ta,tacc,te,tru,ts,tsc,tsg,tsl,tsr,
     .       tu,tuout,tvel,x,nstr,neg,elori,neps)
!*** Calculate displacements, strains and sectional forces
!*****************************************************************
      include 'param.dat'
      dimension coord(mnode+melm,3),el(melm),emate(mmate,mmd),
     .     eqs(melm),eqsout(melm,2),esect(msect,msd),f(mtdf),
     .     ir(melm),ncr(melm),ne(melm,mdim_ne),net(mtdf),
     .     nte(mtdf),pmx(melm),pmy(melm),pn(melm),elori(melm),
     .     ss(melm),t(melm,3,3),ta(melm,mmtx,mmtx),tacc(mtdf),
     .     te(melm,mdim),tru(melm,3,3),ts(melm,mdim),
     .     tsc(melm,mdim),tsg(melm,mdim),tsl(melm,mdim),
     .     tsr(melm,mdim),tu(mtdf),tuout(mnode,3),tvel(mtdf),
     .     x(maxlena),bmatx(mdim,mmtx),de(mdim),dmatx(mdim,mdim),
     .     dul(mmtx),ds(mdim),fd(3),q(3),temp(mdim),tst(mdim,mdim)
!
      if(istage==0.or.nsd==1) then              ! Self-weight
      calculation or static analysis
      if(istage==0.and.neg==0) goto 100         ! If no gravity then no
      self-weight calculation
!
!*** Calculate stiffness equation using the CG method ***
       call cdyna_first(nfree,f,x,neps)
  100 continue
      else
!
!*** Calculation of equation of motion using the CG method ***
       call cdyna(beta_n,dalfa,dbeta,delta_n,dtime,
     .    nfree,f,tacc,tvel,x,neps)
      endif
      do ielm=1,nelm
!
!*** Save obtained displacements ***
       do i=1,2
        do j=1,6
         i1=6*(ne(ielm,i)-1)+j
         temp(j)=x(net(i1))
        enddo
        do k=1,6
         dul(6*(i-1)+k)=temp(k)
        enddo
       enddo
!
```

```
!*** Calculate strain increments and total strain (see Eq. (2.31)) ***
        s=ss(ielm)
       call bmake(ielm,s,bmatx,el,ta,esect,ne)
        do i=1,6
         sum=0.0d0
         do j=1,12
          sum=sum+bmatx(i,j)*dul(j)
         enddo
         de(i)=sum
         te(ielm,i)=te(ielm,i)+sum
        enddo
!
!*** Calculate sectional force increments (see Eq. (2.32)) ***
       call dmake(am,ielm,dmatx,emate,esect,fd,ncr,ne,pmx,pmy,pn,q,ts)
       call trans2(ielm,dul,el,t,tru,tst)
        do i=1,6
         sum=0.0d0
         do j=1,6
          sum=sum+dmatx(i,j)*de(j)
         enddo
         temp(i)=sum
        enddo
         do i=1,6
          sum=0.0d0
          do j=1,6
           sum=sum+tst(i,j)*temp(j)
          enddo
          ds(i)=sum
         enddo
!
!*** Calculate sectional forces for each location in an element (see Eq.
(2.18) and (3.3)) ***
        if(nstr==1) then                ! For truss structures
         j=3
         tsc(ielm,j)=tsc(ielm,j)+ds(j)
         tsg(ielm,j)=tsg(ielm,j)+ds(j)
         tsl(ielm,j)=tsl(ielm,j)+ds(j)
         tsr(ielm,j)=tsr(ielm,j)+ds(j)
        else                            ! For rahmen structures
         do j=3,6
          tsc(ielm,j)=tsc(ielm,j)+ds(j)
          tsg(ielm,j)=tsg(ielm,j)+ds(j)
          tsl(ielm,j)=tsl(ielm,j)+ds(j)
          tsr(ielm,j)=tsr(ielm,j)+ds(j)
         enddo
         gauss=2.0d0/dsqrt(3.0d0)-1.0d0     ! Location of stress
         evaluation point
         if(idnint(s*10.0d0)==0) then
```

```
          gc=gauss/2.0d0
          do j=1,2
           tsc(ielm,j)=tsc(ielm,j)+ds(j)
           tsg(ielm,j)=tsg(ielm,j)+ds(j)-ds(7-j)*el(ielm)*gc
           tsl(ielm,j)=tsl(ielm,j)+ds(j)+ds(7-j)*el(ielm)*0.5d0
           tsr(ielm,j)=tsr(ielm,j)+ds(j)-ds(7-j)*el(ielm)*0.5d0
          enddo
         elseif(idnint(s*10.0d0)==-10) then
          gr=0.5d0-gauss/2.0d0
          do j=1,2
           tsr(ielm,j)=tsr(ielm,j)+ds(j)
           tsc(ielm,j)=tsc(ielm,j)+ds(j)+ds(7-j)*el(ielm)*0.5d0
           tsg(ielm,j)=tsg(ielm,j)+ds(j)+ds(7-j)*el(ielm)*gr
           tsl(ielm,j)=tsl(ielm,j)+ds(j)+ds(7-j)*el(ielm)
          enddo
         elseif(idnint(s*10.0d0)==10) then
          gl=0.5d0+gauss/2.0d0
          do j=1,2
           tsl(ielm,j)=tsl(ielm,j)+ds(j)
           tsc(ielm,j)=tsc(ielm,j)+ds(j)-ds(7-j)*el(ielm)*0.5d0
           tsg(ielm,j)=tsg(ielm,j)+ds(j)-ds(7-j)*el(ielm)*gl
           tsr(ielm,j)=tsr(ielm,j)+ds(j)-ds(7-j)*el(ielm)
          enddo
         elseif(idint(s*10.0d0)==-1) then
          gc=gauss/2.0d0
          gl=0.5d0+gauss/2.0d0
          gr=0.5d0-gauss/2.0d0
          do j=1,2
           tsg(ielm,j)=tsg(ielm,j)+ds(j)
           tsc(ielm,j)=tsc(ielm,j)+ds(j)+ds(7-j)*el(ielm)*gc
           tsl(ielm,j)=tsl(ielm,j)+ds(j)+ds(7-j)*el(ielm)*gl
           tsr(ielm,j)=tsr(ielm,j)+ds(j)-ds(7-j)*el(ielm)*gr
          enddo
         else
          write(ii9,'(21hss(ielm) ERROR: ielm=,i5,4h s=,e11.4)') ielm,s
          stop
         endif
        endif
!
!*** Calculate yield function (see Eq. 3.4) ***
        if(nmetd==1) then                ! For conveitional FEM
         eqs(ielm)=(tsc(ielm,1)/pmx(ielm))*(tsc(ielm,1)/pmx(ielm))
     .        +(tsc(ielm,2)/pmy(ielm))*(tsc(ielm,2)/pmy(ielm))
     .        +(tsc(ielm,3)/pn(ielm))*(tsc(ielm,3)/pn(ielm))
         eqsout(ielm,1)=eqs(ielm)        ! For graphics data (values at mid-
                                         ! points of elements)
         eqsout(ielm,2)=eqs(ielm)
```

```
          if(eqsout(ielm,1)>1.0d0) eqsout(ielm,1)=1.0d0
          if(eqsout(ielm,2)>1.0d0) eqsout(ielm,2)=1.0d0
        else                               ! For ASI and ASI-Gauss
       techniques
          eqsl=(tsl(ielm,1)/pmx(ielm))*(tsl(ielm,1)/pmx(ielm))
     .      +(tsl(ielm,2)/pmy(ielm))*(tsl(ielm,2)/pmy(ielm))
     .      +(tsl(ielm,3)/pn(ielm))*(tsl(ielm,3)/pn(ielm))
                                           ! Values at mid-points of members
          eqsr=(tsr(ielm,1)/pmx(ielm))*(tsr(ielm,1)/pmx(ielm))
     .      +(tsr(ielm,2)/pmy(ielm))*(tsr(ielm,2)/pmy(ielm))
     .      +(tsr(ielm,3)/pn(ielm))*(tsr(ielm,3)/pn(ielm))
                                           ! Values at either end of members
          eqsout(ielm,1)=eqsl              ! For graphics data (values at
       both ends of elements)
          eqsout(ielm,2)=eqsr
          if(eqsl>1.0d0) eqsout(ielm,1)=1.0d0
          if(eqsr>1.0d0) eqsout(ielm,2)=1.0d0
          if(ncr(ielm)<=0) then
                                          ! If in elastic range, larger
          value between eqsl and eqsr is assigned as eqs
           ir(ielm)=0
            if(eqsl>eqsr) then
            eqs(ielm)=eqsl
            if(eqsl>=1.0d0) ir(ielm)=-1
           elseif(eqsr>=eqsl) then
            eqs(ielm)=eqsr
            if(eqsr>=1.0d0) ir(ielm)=1
           endif
          elseif(ncr(ielm)>0) then ! If in plastic range
           if(ir(ielm)==-1) eqs(ielm)=eqsl
           if(ir(ielm)==1) eqs(ielm)=eqsr
          endif
        endif
!
!*** Classification of elastic/yielding/unloading ***
        if(ncr(ielm)<=0) then              ! If in elastic range
          if(eqs(ielm)>=1.0d0) then
           ncr(ielm)=1                     ! If f_y > 1.0 then the element is
          yielded (ncr=1)
          else
           if(ncr(ielm)<0) ncr(ielm)=-2 ! If still unloaded
           (ncr=-2) cycle
          endif
        elseif(ncr(ielm)>0) then           ! If the element is yielded
          ul=0.0d0
          do i=1,3
           ul=ul+q(i)*de(i)
          enddo
```

```
        ul=ul/am
        if(ul<0.0d0) ncr(ielm)=-1          ! If unloaded at this
        step (ncr=-1)
        if(ul>=0.0d0) ncr(ielm)=2          ! If still yielded (ncr=2)
       endif
       do j=1,6                            ! Sectional forces resized
       due to plastic theory
        tsc(ielm,j)=tsc(ielm,j)/dsqrt(eqs(ielm))
        tsg(ielm,j)=tsg(ielm,j)/dsqrt(eqs(ielm))
        tsl(ielm,j)=tsl(ielm,j)/dsqrt(eqs(ielm))
        tsr(ielm,j)=tsr(ielm,j)/dsqrt(eqs(ielm))
       enddo
       eqs(ielm)=1.0d0
      enddo
!
!*** Calculate displacements, node coordinates and element lengths ***
      do i=1,nfree
       tu(nte(i))=tu(nte(i))+x(i)
      enddo
      do inode=1,node
       do j=1,3
        tuout(inode,j)=coord(inode,j)+tu(6*(inode-1)+j)
       enddo
      enddo
      do ielm=1,nelm
       xx=tuout(ne(ielm,2),1)-tuout(ne(ielm,1),1)
       yy=tuout(ne(ielm,2),2)-tuout(ne(ielm,1),2)
       zz=tuout(ne(ielm,2),3)-tuout(ne(ielm,1),3)
       el(ielm)=dsqrt(xx*xx+yy*yy+zz*zz)
      enddo
!
      return
      end
!*******************************************************************
```

⋆Explanations on variables and arrays

`te(melm,mdim)`: Total strain

`de(mdim)`: Strain increment

`t(melm,3,3)`: Components of single transformation matrix

`tru(melm,3,3)`: Components of transformation matrix using ULF

`dul(mmtx)`: Displacement increment at each DOF

`ds(mdim)`: Sectional force increment

`tst(mdim,mdim)`: Components of the matrix for transforming updated Kirchhoff sectional force increment vector to Jaumann differential form vector

A.3.16 Subroutine "tmasm"

```
!*****************************************************************
      subroutine tmasm(nelm,nmm,nfree,ntdf,elori,emate,esect,ne,ta)
!*** Make total mass matrix
!*****************************************************************
      include 'param.dat'
      dimension elori(melm),emate(mmate,mmd),esect(msect,msd),
     .    ne(melm,mdim_ne),ta(melm,mmtx,mmtx),
     .    emmat(mmtx,mmtx)
!
      call gminit(nfree,ntdf-nfree)
!
!*** Make elemental mass matrix and total mass matrix ***
      do ielm=1,nelm
        call elmass(ielm,nmm,elori,emate,emmat,esect,ne,ta)
        call insmmat(ielm,emmat,ke,icon)
      enddo
!
      return
      end
!*****************************************************************
```

*Comments

This subroutine makes total mass matrix for the CG method. The subroutines "gminit" and "insmmat" are omitted in this appendix.

A.3.17 Subroutine "transl"

```
!*****************************************************************
      subroutine transl(istage,nelm,coord,el,elori,ne,t,ta,tru,nerr)
!*** Make initial transformation matrix
!*****************************************************************
      include 'param.dat'
      dimension coord(mnode+melm,3),el(melm),elori(melm),
     .     ne(melm,mdim_ne),t(melm,3,3),ta(melm,mmtx,mmtx),
     .     tru(melm,3,3),dji(3),dki(3),temp(3,3)
!
      if(istage.eq.0.and.nerr.eq.1)
     .write(ii9,*) '-- Searching for input errors --'
!
      if(istage.eq.0) then
       do ielm=1,nelm
        ip=ne(ielm,1)
        jp=ne(ielm,2)
        kp=ne(ielm,3)
!
      if(nerr.eq.1) write(ii9,*) 'Element No. = ',ielm
```

```
!
!*** Calculate element lengths ***
        xx=coord(jp,1)-coord(ip,1)
        yy=coord(jp,2)-coord(ip,2)
        zz=coord(jp,3)-coord(ip,3)
        el(ielm)=dsqrt(xx*xx+yy*yy+zz*zz)
        elori(ielm)=el(ielm)
!
!*** Calculate components of initial transformation matrix using
reference nodes ***
        do j=1,3
         dji(j)=coord(jp,j)-coord(ip,j)
         dki(j)=coord(kp,j)-coord(ip,j)
        enddo
        vp1=dji(2)*dki(3)-dji(3)*dki(2)
        vp2=dji(3)*dki(1)-dji(1)*dki(3)
        vp3=dji(1)*dki(2)-dji(2)*dki(1)
        vp=dsqrt(vp1*vp1+vp2*vp2+vp3*vp3)
        t(ielm,3,1)=dji(1)/elori(ielm)
        t(ielm,3,2)=dji(2)/elori(ielm)
        t(ielm,3,3)=dji(3)/elori(ielm)
        t(ielm,2,1)=vp1/vp
        t(ielm,2,2)=vp2/vp
        t(ielm,2,3)=vp3/vp
        t(ielm,1,1)=t(ielm,2,2)*t(ielm,3,3)-t(ielm,2,3)*t(ielm,3,2)
        t(ielm,1,2)=t(ielm,2,3)*t(ielm,3,1)-t(ielm,2,1)*t(ielm,3,3)
        t(ielm,1,3)=t(ielm,2,1)*t(ielm,3,2)-t(ielm,2,2)*t(ielm,3,1)
        do j=1,mmtx
         do k=1,mmtx
          ta(ielm,j,k)=0.0d0
         enddo
        enddo
       if(nerr.eq.1) write(ii9,*) ' O.K.'
       enddo
!
!*** Update transformation matrix using ULF (see Eq. (2.26)) ***
      elseif(istage>0) then
       do ielm=1,nelm
        do i=1,3
         do j=1,3
          temp(i,j)=t(ielm,i,j)
         enddo
        enddo
        do i=1,3
         do j=1,3
          sum=0.0d0
```

```
          do k=1,3
           sum=sum+tru(ielm,i,k)*temp(k,j)
          enddo
          t(ielm,i,j)=sum
         enddo
        enddo
       enddo
      endif
!*** Make transformation matrix for 12 DOF (see Eq. (2.28)) ***
      do ielm=1,nelm
       do j=1,3
        do k=1,3
         ta(ielm,j,k)=t(ielm,j,k)
         ta(ielm,j+3,k+3)=t(ielm,j,k)
         ta(ielm,j+6,k+6)=t(ielm,j,k)
         ta(ielm,j+9,k+9)=t(ielm,j,k)
        enddo
       enddo
      enddo
!
      return
      end
!*******************************************************************
```

★Explanations on variables and arrays

`dji(3)`: Vector along the elemental axis

`dki(3)`: Vector from the starting node (s=-1) to reference node of the element

`t(i,3,j)`: Unit vector of dji(i) (z direction in elemental coordinate system)

`t(i,2,j)`: Unit vector of {vp} (y direction in elemental coordinate system)

`t(i,1,j)`: Unit vector vertical to the elemental axis (x direction in elemental coordinate system)

A.3.18 Subroutine "`trans2`"

```
!*******************************************************************
      subroutine trans2(ielm,dul,el,t,tru,tst)
!*** Make transformation matrix for ULF
!*******************************************************************
      include 'param.dat'
      dimension dul(mmtx),el(melm),t(melm,3,3),tru(melm,3,3),
     .     tst(mdim,mdim),du(3)
!
```

```
!*** Calculate relative nodal displacement increments in elemental
coordinate system ***
      do i=1,3
       sum=0.0d0
       do j=1,3
        sum=sum+t(ielm,i,j)*(dul(j+6)-dul(j))
       enddo
       du(i)=sum
      enddo
!
!*** Calculate sinα, cosα, sinβ, cosβ, sinγ and cosγ (see Eq. (2.30)) ***
      dz=el(ielm)+du(3)
      sina=-du(2)/dsqrt(dz*dz+du(2)*du(2))
      cosa=dz/dsqrt(dz*dz+du(2)*du(2))
      sinb=du(1)/dsqrt(dz*dz+du(2)*du(2)+du(1)*du(1))
      cosb=dsqrt(dz*dz+du(2)*du(2))/dsqrt(dz*dz+du(2)*du(2)
     .   +du(1)*du(1))
      sum=0.0d0
      do j=1,3
       sum=sum+t(ielm,3,j)*(dul(j+3)+dul(j+9))
      enddo
      cosc=dcos(sum/2.0d0)
      sinc=dsin(sum/2.0d0)
!
!*** Calculate transformation matrix (see Eq. (2.29)) ***
      tru(ielm,1,1)=cosb*cosc
      tru(ielm,1,2)=sina*sinb*cosc+cosa*sinc
      tru(ielm,1,3)=-cosa*sinb*cosc+sina*sinc
      tru(ielm,2,1)=-cosb*sinc
      tru(ielm,2,2)=-sina*sinb*sinc+cosa*cosc
      tru(ielm,2,3)=cosa*sinb*sinc+sina*cosc
      tru(ielm,3,1)=sinb
      tru(ielm,3,2)=-sina*cosb
      tru(ielm,3,3)=cosa*cosb
!
!*** Make matrix for transforming updated Kirchhoff sectional force
increment vector to Jaumann differential form vector (see Eq. (2.34)) ***
      do i=1,mdim
       do j=1,mdim
        tst(i,j)=0.0d0
       enddo
      enddo
      tst(1,1)=tru(ielm,1,1)
      tst(1,2)=tru(ielm,1,2)
      tst(1,4)=tru(ielm,1,3)
      tst(2,1)=tru(ielm,2,1)
      tst(2,2)=tru(ielm,2,2)
```

```
      tst(2,4)=tru(ielm,2,3)
      tst(3,3)=tru(ielm,3,3)
      tst(3,5)=tru(ielm,3,1)
      tst(3,6)=tru(ielm,3,2)
      tst(4,1)=tru(ielm,3,1)
      tst(4,2)=tru(ielm,3,2)
      tst(4,4)=tru(ielm,3,3)
      tst(5,3)=tru(ielm,1,3)
      tst(5,5)=tru(ielm,1,1)
      tst(5,6)=tru(ielm,1,2)
      tst(6,3)=tru(ielm,2,3)
      tst(6,5)=tru(ielm,2,1)
      tst(6,6)=tru(ielm,2,2)
!
      return
      end
!*****************************************************************
```

★Explanations on variables and arrays

`du(3)`: Relative nodal displacement increments in elemental coordinate system

A.3.19 Subroutine "tstfns"

```
!*****************************************************************
      subroutine tstfns(nelm,nfree,ntdf,el,elori,emate,esect,ncr,ne,
     .        pmx,pmy,pn,ss,ta,ts)
!*** Make total stiffness matrix

!*****************************************************************
      include 'param.dat'
      dimension el(melm),elori(melm),emate(mmate,mmd),
     .   esect(msect,msd),ncr(melm),ne(melm,mdim_ne),
     .   pn(melm),ss(melm),ta(melm,mmtx,mmtx),ts(melm,mdim),
     .   ekmat(mmtx,mmtx),pmx(melm),pmy(melm)
!
!*** Initialize total stiffness matrix ***
      call gkinit(nfree,ntdf-nfree)
!
!*** Make elemental stiffness matrix and total stiffness matrix ***
      do ielm=1,nelm
       call elstif(ielm,ekmat,el,elori,emate,esect,ncr,ne,pmx,
     .   pmy,pn,ss,ta,ts)
       call inskmat(ielm,ekmat,ke,icon)
      enddo
!
      return
      end
!*****************************************************************
```

*Comments

This subroutine makes total stiffness matrix for the CG method. The subroutines "gkinit" and "inskmat" are omitted in this appendix.

A.3.20 Include File "param.dat"

```
!*****************************************************************
      implicit Double Precision(a-h,o-z)
      parameter(melm=1000,mtdf=6000,mnode=1000,mnc=300,mmate=15,
     .      msect=15,mdim=6,mdim_ne=12,mmtx=12,mmd=5,msd=15,
     .      ii4=70,ii5=15,ii6=16,ii7=17,ii9=19,
     .      nofpn=6,nonpe=2,noef=nofpn*nonpe,
     .      maxlenk=98000,maxnode=1200,maxfr=nofpn*maxnode,
     .      maxlena=5000,maxlenb=maxfr-maxlena)
      common /gk/gkmat(maxlenk),gmmat(maxlenk),jtab(maxlenk)
      common /id/ilm(maxfr),irma(maxfr),irmb(maxfr)
      common /klux/ke(2,melm),icon(6,mnode)
!*****************************************************************
```

*Comments

This file is included in all subroutines and the main program. Maximum numbers of elements, nodes, DOFs and constraint DOFs can be changed according to the size of the numerical model.

A.4 SUMMARY

The author and his collaborators are happy to receive comments on the numerical code, however, they have no responsibility of whatsoever caused from the results of using the code. The numerical code and user's manuals uploaded on the website are to be revised without notice.

The files uploaded on the website contain:
asifem_v1_3.exe: Numerical code execution file (ASIFEM Ver.1.3)
input.txt: Input data file
output.txt, post.out, eqs.out: Output data files
manual_e.pdf: User's manual
Graphic.exe: Graphic software execution file (using post.out, eqs.out as input files)

`color_range.txt`: Color range file for the graphic software
`glut32.dll`: Dynamic library for the graphic software
`manual(Graphic.exe)_e.txt`: User's manual for the graphic software

REFERENCE

Website of ASIFEM: Finite element code for framed structures using ASI-Gauss technique. http://www.kz.tsukuba.ac.jp/~isobe/asifem_e.html.

APPENDIX B

ASI Technique Utilizing Bernoulli–Euler Beam Elements

B.1 INTRODUCTION

As described in Chapter 2, there are two versions of the original ASI technique (Toi and Isobe, 1993, 1996, 2000): one utilizing Bernoulli–Euler beam elements and another utilizing linear Timoshenko beam elements. Although fractured section on either sides of a member cannot be considered by a single Bernoulli–Euler beam element, the bending deformations in the element can be calculated with high precision using a two-point Gaussian quadrature. In this appendix, an incremental finite-element formulation of the Bernoulli–Euler beam element followed by a basic theory of the ASI technique, including a shifting technique of numerical integration points and their relations to stress evaluation points, are described. Furthermore, some numerical examples are shown to confirm the validity of the technique.

B.2 BERNOULLI–EULER BEAM ELEMENT

The relation between the generalized strain increment and nodal displacement increment vectors of a Bernoulli–Euler beam element (known as a cubic beam element in other descriptions) are given, as in Eq. (2.2), by the following equation:

$$\{\Delta_n \varepsilon(r)\} = [B(s)]\{\Delta_n u\} \tag{B.1}$$

where s is the location of the numerical integration point and r is the location of the stress evaluation point where stresses and strains are actually evaluated. Other nomenclatures follow those described in other chapters. The components of the generalized strain increment vector $\{\Delta_n \varepsilon(r)\}$ at incremental step n are varied from Eq. (2.3) and are given by the following equations:

$$\Delta_n\kappa_x(r) = \frac{d^2\Delta_n v(z)}{dz^2} \tag{B.2a}$$

$$\Delta_n\kappa_y(r) = \frac{d^2\Delta_n u(z)}{dz^2} \tag{B.2b}$$

$$\Delta_n\varepsilon_z(r) = \frac{d\Delta_n w(z)}{dz} \tag{B.2c}$$

$$\Delta_n\theta_{z,z}(r) = \frac{d\Delta_n\theta_z(z)}{dz} \tag{B.2d}$$

where the displacement components (u, v, w, θ_z) are defined as shown in Fig. 2.1. In the Bernoulli–Euler beam element, the displacement functions for bending deformations are defined by cubic functions, and those for axial and torsional deformations are defined by linear functions as follows:

$$u(z) = H_{00}u_1 + H_{10}{}^n l\theta_{y1} + H_{01}u_2 + H_{11}{}^n l\theta_{y2} \tag{B.3a}$$

$$v(z) = H_{00}v_1 - H_{10}{}^n l\theta_{x1} + H_{01}v_2 - H_{11}{}^n l\theta_{x2} \tag{B.3b}$$

$$w(z) = \frac{1}{2}(1-s)w_1 + \frac{1}{2}(1+s)w_2 \tag{B.3c}$$

$$\theta_z(z) = \frac{1}{2}(1-s)\theta_{z1} + \frac{1}{2}(1+s)\theta_{z2} \tag{B.3d}$$

where

$$H_{00} = \frac{1}{8}\left(4 - 6s + 2s^3\right) \tag{B.4a}$$

$$H_{10} = \frac{1}{8}\left(1 - s - s^2 + s^3\right) \tag{B.4b}$$

$$H_{01} = \frac{1}{8}\left(4 + 6s - 2s^3\right) \tag{B.4c}$$

$$H_{11} = \frac{1}{8}\left(-1 - s + s^2 + s^3\right) \tag{B.4d}$$

The subscripts 1 and 2 denote the node numbers indicated in Fig. 2.1. By substituting Eq. (B.3) into Eq. (B.2), one can obtain the relations between the generalized strain increments and nodal displacement increments as follows:

$$\Delta_n\kappa_x(r) = \frac{6s}{{}^{n}l^2}\Delta_n v_1 - \frac{1}{{}^{n}l}(3s-1)\Delta_n\theta_{x1} - \frac{6s}{{}^{n}l^2}\Delta_n v_2 - \frac{1}{{}^{n}l}(3s+1)\Delta_n\theta_{x2} \tag{B.5a}$$

$$\Delta_n\kappa_y(r) = \frac{6s}{{}^{n}l^2}\Delta_n u_1 + \frac{1}{{}^{n}l}(3s-1)\Delta_n\theta_{y1} - \frac{6s}{{}^{n}l^2}\Delta_n u_2 + \frac{1}{{}^{n}l}(3s+1)\Delta_n\theta_{y2} \tag{B.5b}$$

$$\Delta_n\varepsilon_z(r) = \frac{1}{{}^{n}l}(\Delta_n w_2 - \Delta_n w_1) \tag{B.5c}$$

$$\Delta_n\theta_{z,z}(r) = \frac{1}{{}^{n}l}(\Delta_n\theta_{z2} - \Delta_n\theta_{z1}) \tag{B.5d}$$

In summary, the components of the generalized strain increment vector $\{\Delta_n\varepsilon(r)\}$, the nodal displacement increment vector $\{\Delta_n u\}$, and the generalized strain–nodal displacement matrix $[B(s)]$ of Eq. (B.1) are given as follows:

$$\{\Delta_n\varepsilon(r)\} = \left\{\Delta_n\kappa_x(r) \quad \Delta_n\kappa_y(r) \quad \Delta_n\varepsilon_z(r) \quad \Delta_n\theta_{z,z}(r)\right\}^T \tag{B.6}$$

$$\begin{aligned}\{\Delta_n u\} = \{&\Delta_n u_1 \quad \Delta_n v_1 \quad \Delta_n w_1 \quad \Delta_n\theta_{x1} \quad \Delta_n\theta_{y1} \quad \Delta_n\theta_{z1} \quad \Delta_n u_2 \quad \Delta_n v_2 \\ &\Delta_n w_2 \quad \Delta_n\theta_{x2} \quad \Delta_n\theta_{y2} \quad \Delta_n\theta_{z2}\}^T\end{aligned} \tag{B.7}$$

$$[B(s)] = \begin{bmatrix} 0 & \frac{6s}{{}^{n}l^2} & 0 & -\frac{3s-1}{{}^{n}l} & 0 & 0 & 0 & -\frac{6s}{{}^{n}l^2} & 0 & -\frac{3s+1}{{}^{n}l} & 0 & 0 \\ \frac{6s}{{}^{n}l^2} & 0 & 0 & 0 & \frac{3s-1}{{}^{n}l} & 0 & -\frac{6s}{{}^{n}l^2} & 0 & 0 & 0 & \frac{3s+1}{{}^{n}l} & 0 \\ 0 & 0 & -\frac{1}{{}^{n}l} & 0 & 0 & 0 & 0 & 0 & \frac{1}{{}^{n}l} & 0 & 0 & 0 \\ 0 & 0 & 0 & 0 & 0 & -\frac{1}{{}^{n}l} & 0 & 0 & 0 & 0 & 0 & \frac{1}{{}^{n}l} \end{bmatrix} \tag{B.8}$$

At those phases when the element behaves elastically, the relation between generalized sectional force increment vector and generalized strain increment vector is given by the following equation:

$$\{\Delta_n\sigma(r)\} = [D_e(r)]\{\Delta_n\varepsilon(r)\} \tag{B.9}$$

where $[D_e(r)]$ is the generalized sectional force–generalized strain matrix for elastic deformation.

The components of the generalized sectional force increment vector $\{\Delta_n\sigma(r)\}$ at incremental step n for a Bernoulli–Euler beam element decrease in number from Eq. (2.10), and are given by the following equations:

$$\Delta_n M_x(r) = EI_x \Delta_n \kappa_x(r) \tag{B.10a}$$

$$\Delta_n M_y(r) = EI_y \Delta_n \kappa_y(r) \tag{B.10b}$$

$$\Delta_n N(r) = EA \Delta_n \varepsilon_z(r) \tag{B.10c}$$

$$\Delta_n M_z(r) = GK \Delta_n \theta_{z,z}(r) \tag{B.10d}$$

In summary, the components of the generalized sectional force increment vector $\{\Delta_n\sigma(r)\}$ and the generalized sectional force–generalized strain matrix $[D_e(r)]$ of Eq. (B.9) can be given as follows:

$$\{\Delta_n\sigma(r)\} = \{\Delta_n M_x(r) \quad \Delta_n M_y(r) \quad \Delta_n N(r) \quad \Delta_n M_z(r)\}^T \tag{B.11}$$

$$[D_e(r)] = \begin{bmatrix} EI_x & 0 & 0 & 0 \\ 0 & EI_y & 0 & 0 \\ 0 & 0 & EA & 0 \\ 0 & 0 & 0 & GK \end{bmatrix} \tag{B.12}$$

It is preferable, also for this version of the ASI technique, to use the distributed elemental mass matrix for dynamic analyses, which is given for the Bernoulli–Euler beam element as the following equation:

$$[M_E] = \frac{\rho^n l}{420}\begin{bmatrix}
156A & 0 & 0 & 0 & 22A^n l & 0 & 54A & 0 & 0 & 0 & -13A^n l & 0 \\
0 & 156A & 0 & -22A^n l & 0 & 0 & 0 & 54A & 0 & 13A^n l & 0 & 0 \\
0 & 0 & 140A & 0 & 0 & 0 & 0 & 0 & 70A & 0 & 0 & 0 \\
0 & -22A^n l & 0 & 4A^n l^2 & 0 & 0 & 0 & -13A^n l & 0 & -3A^n l^2 & 0 & 0 \\
22A^n l & 0 & 0 & 0 & 4A^n l^2 & 0 & 13A^n l & 0 & 0 & 0 & -3A^n l^2 & 0 \\
0 & 0 & 0 & 0 & 0 & 140I_z & 0 & 0 & 0 & 0 & 0 & 70I_z \\
54A & 0 & 0 & 0 & 13A^n l & 0 & 156A & 0 & 0 & 0 & -22A^n l & 0 \\
0 & 54A & 0 & -13A^n l & 0 & 0 & 0 & 156A & 0 & 22A^n l & 0 & 0 \\
0 & 0 & 70A & 0 & 0 & 0 & 0 & 0 & 140A & 0 & 0 & 0 \\
0 & 13A^n l & 0 & -3A^n l^2 & 0 & 0 & 0 & 22A^n l & 0 & 4A^n l^2 & 0 & 0 \\
-13A^n l & 0 & 0 & 0 & -3A^n l^2 & 0 & -22A^n l & 0 & 0 & 0 & 4A^n l^2 & 0 \\
0 & 0 & 0 & 0 & 0 & 70I_z & 0 & 0 & 0 & 0 & 0 & 140I_z
\end{bmatrix} \tag{B.13}$$

B.3 ASI TECHNIQUE UTILIZING BERNOULLI–EULER BEAM ELEMENT

In the ASI technique utilizing Bernoulli–Euler beam element, initial locations of the numerical integration points, when the element acts elastically, are Gaussian integration points used for two-point quadrature. If a fully plastic section is determined in an element, the numerical integration points are shifted immediately after the determination in order to express plastic hinges, in the same manner as in other ASI techniques.

Fig. B.1 shows a Bernoulli–Euler beam element and its physical equivalence to a rigid-body-spring-model (RBSM). The relationship between the locations of the numerical integration and stress evaluation points where plastic hinges are actually formed is expressed as follows (Toi, 1991):

$$s_1 = \frac{1}{3r_1} = -\frac{1}{3r_2}, \quad s_2 = -\frac{1}{3r_1} = \frac{1}{3r_2} \quad (r_1 = -r_2) \tag{B.14}$$

where (s_1, s_2) and (r_1, r_2) are the locations of numerical integration and stress evaluation points, respectively. As mentioned, the numerical integration points are initially placed at the Gaussian integration points of the element ($s_i = \mp 1/\sqrt{3}$; $i = 1,2$), which gives $r_i = \mp 1/\sqrt{3}$ ($i = 1,2$) from Eq. (B.14), when the entire region of an element behaves elastically. Therefore, the elemental stiffness matrix for the element in elastic range is given by the following equation:

$$\begin{aligned}[K_E] = \frac{l}{2}\Bigg\{&\left[B\left(-\frac{1}{\sqrt{3}}\right)\right]^T\left[D_e\left(-\frac{1}{\sqrt{3}}\right)\right]\left[B\left(-\frac{1}{\sqrt{3}}\right)\right] \\ &+\left[B\left(\frac{1}{\sqrt{3}}\right)\right]^T\left[D_e\left(\frac{1}{\sqrt{3}}\right)\right]\left[B\left(\frac{1}{\sqrt{3}}\right)\right]\Bigg\}\end{aligned} \tag{B.15}$$

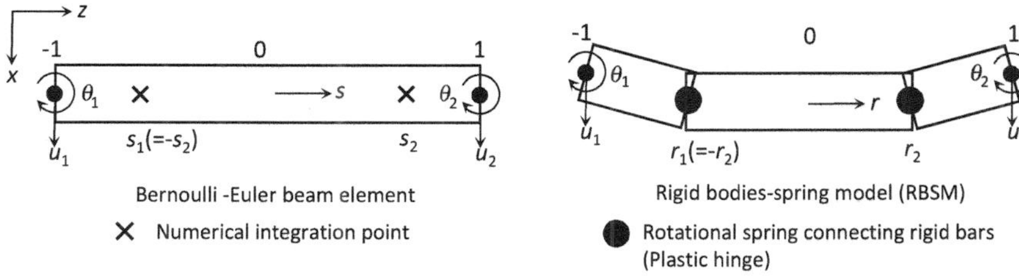

Figure B.1 Bernoulli–Euler beam element and its physical equivalence to RBSM.

The generalized strain increment vector and the generalized sectional force increment vector are calculated as follows:

$$\left\{\Delta_n\varepsilon\left(\pm\frac{1}{\sqrt{3}}\right)\right\} = \left[B\left(\pm\frac{1}{\sqrt{3}}\right)\right]\{\Delta_n u\} \tag{B.16}$$

$$\left\{\Delta_n\sigma\left(\pm\frac{1}{\sqrt{3}}\right)\right\} = \left[D_e\left(\pm\frac{1}{\sqrt{3}}\right)\right]\left\{\Delta_n\varepsilon\left(\pm\frac{1}{\sqrt{3}}\right)\right\} \tag{B.17}$$

The bending moments calculated from Eq. (B.17) are physically those at the Gaussian integration points. Thus, the bending moments along an elastically deformed element can be determined by the following approximations:

$$\Delta_n M_x(s) = \frac{1}{2}\Delta_n M_x\left(-\frac{1}{\sqrt{3}}\right)\left(1-\sqrt{3}\,s\right) + \frac{1}{2}\Delta_n M_x\left(\frac{1}{\sqrt{3}}\right)\left(1+\sqrt{3}\,s\right) \tag{B.18a}$$

$$\Delta_n M_y(s) = \frac{1}{2}\Delta_n M_y\left(-\frac{1}{\sqrt{3}}\right)\left(1-\sqrt{3}\,s\right) + \frac{1}{2}\Delta_n M_y\left(\frac{1}{\sqrt{3}}\right)\left(1+\sqrt{3}\,s\right) \tag{B.18b}$$

After the formation of a fully plastic section at either one of the ends of the element ($r = r_i = \mp 1$), the numerical integration points are shifted immediately to new points ($s = s_i = \mp 1/3$, according to Eq. (B.14)). For example, if a fully plastic section first occurs at the right end of an element ($r = r_2 = 1$), the elemental stiffness matrix, the generalized strain, and sectional force increment vectors are given as follows:

$$\begin{aligned}[K_E] = \frac{{}^n l}{2}\Bigg\{&\left[B\left(-\frac{1}{3r_2}\right)\right]^T [D_e(-r_2)]\left[B\left(-\frac{1}{3r_2}\right)\right] \\ &+ \left[B\left(\frac{1}{3r_2}\right)\right]^T [D_p(r_2)]\left[B\left(\frac{1}{3r_2}\right)\right]\Bigg\}\end{aligned} \tag{B.19a}$$

$$\{\Delta_n\varepsilon(\pm r_2)\} = \left[B\left(\pm\frac{1}{3r_2}\right)\right]\{\Delta_n u\} \tag{B.19b}$$

$$\{\Delta_n\sigma(r_2)\} = [D_p(r_2)]\{\Delta_n\varepsilon(r_2)\} \tag{B.19c}$$

$$\{\Delta_n\sigma(-r_2)\} = [D_e(-r_2)]\{\Delta_n\varepsilon(-r_2)\} \tag{B.19d}$$

where $[D_p(r)]$ is the generalized sectional force–generalized strain matrix for plastic deformation. The sectional force increments calculated by

Eqs. (B.19c) and (B.19d) are physically those at $r = \pm r_2$, which, therefore, should be added to the sectional forces calculated at $r = \pm r_2$ before yielding occurred. After another fully plastic section occurs at the other end of the element ($r = r_1 = -r_2 = -1$), the generalized sectional force–generalized strain matrix $[D_e(-r_2)]$ in Eqs. (B.19a) and (B.19d) is replaced by $[D_p(-r_2)]$. The numerical integration points will return to their initial locations, if unloading occurs.

B.4 NUMERICAL EXAMPLES

In this section, some numerical examples of elastic and elasto-plastic analyses of a simple space frame are discussed (Toi and Isobe, 1993). The space frame shown in Fig. 3.2 was subjected to static and dynamic horizontal loads. In these analyses, the following two methods, both utilizing Bernoulli–Euler beam elements, were applied to verify the accuracies of each method: (1) conventional finite-element (FEM) in which the numerical integration points of each element are fixed at their initial locations ($s = \pm 1/\sqrt{3}$) throughout the analysis; and (2) FEM using the ASI technique.

Fig. B.2 shows the numerical results of the elastic analyses under static loading. As the two-point Gaussian quadrature is used in the beam elements,

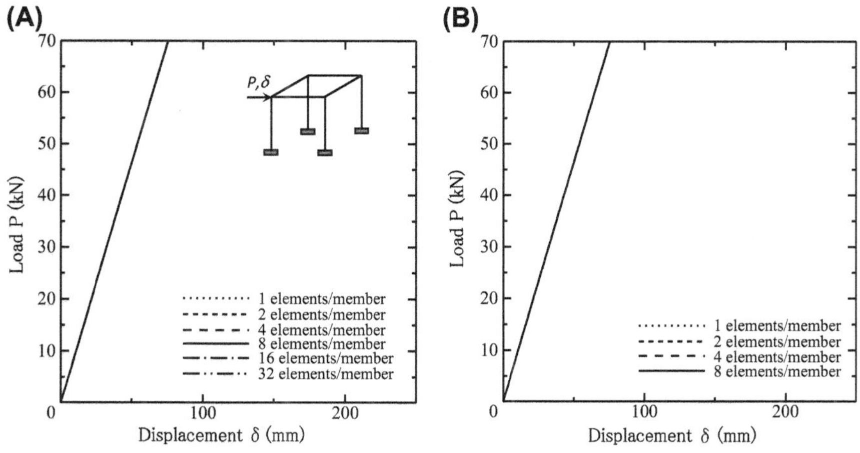

Figure B.2 Load–displacement relations in elastic analysis. (A) Conventional FEM and (B) ASI technique.

exact elastic solutions were obtained with a minimum subdivision of one element per member by both methods. However, as shown in Fig. B.3, the differences in the plastic collapse loads clearly appear in the results of the elasto-plastic analysis. The convergence is extremely slow for the conventional FEM as observed in the case of the linear Timoshenko beam element in Chapter 3. About eight elements per member were required to obtain the exact plastic collapse load. On the other hand, highly accurate solutions were obtained by the ASI technique, and only one-element subdivision per member was required to obtain the converged solution.

Figs. B.4 and B.5 show the numerical results of the elastic response and elasto-plastic response analyses, where the same tendencies as static analyses can be observed. There is no problem in analyzing the behaviors by conventional FEM with less-element subdivisions only if the loading range is elastic. The differences appear in elasto-plastic response analysis, where the locations of plastic hinges tend to affect the responses largely. The solutions by the ASI technique utilizing the Bernoulli–Euler beam element tend to become slightly deformable compared to the exact solution, however, the result shows fast convergence also in the dynamic analyses. The modeling with smaller number of elements can practically give sufficient solutions in problems where vibration modes of higher order are not the main issue to be discussed.

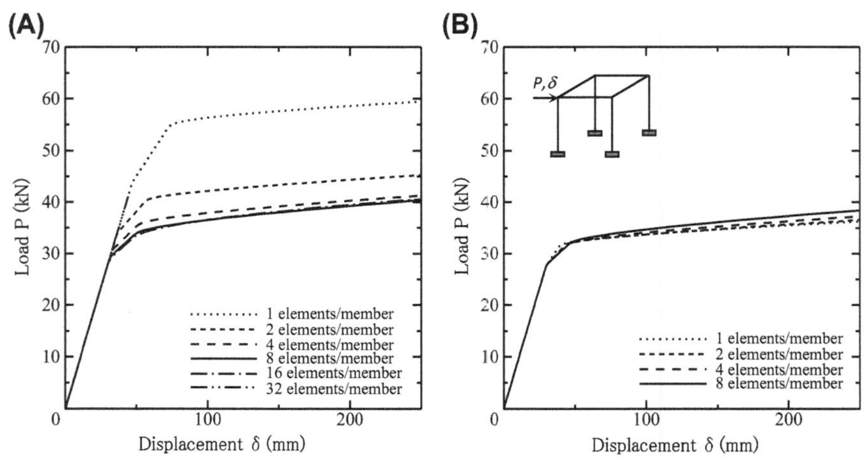

Figure B.3 Load–displacement relations in elasto-plastic analysis. (A) Conventional FEM and (B) ASI technique.

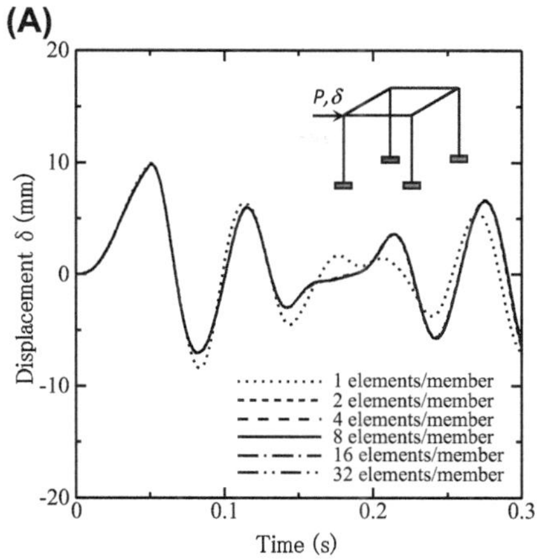

Figure B.4 Elastic responses of a simple space frame. (A) Conventional FEM and (B) ASI technique.

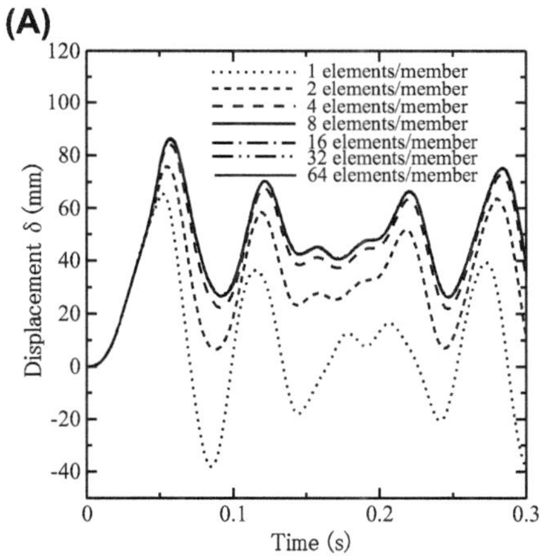

Figure B.5 Elasto-plastic responses of a simple space frame. (A) Conventional FEM and (B) ASI technique.

B.5 SUMMARY

Basic theories of the ASI technique utilizing Bernoulli–Euler beam elements are described in this appendix. This technique cannot be applied to collapse problems, however, it requires only one element per member for the modeling and it is very easy to simulate and understand simple behaviors of framed structures. The code may be used in basic education for structural engineers and students, and it can also be downloaded from the author's website (Website of ASIFEM3).

REFERENCES

Toi, Y., 1991. Shifted integration technique in one-dimensional plastic collapse analysis using linear and cubic finite elements. International Journal for Numerical Methods in Engineering 31, 1537–1552.

Toi, Y., Isobe, D., 1993. Adaptively shifted integration technique for finite element collapse analysis of framed structures. International Journal for Numerical Methods in Engineering 36, 2323–2339.

Toi, Y., Isobe, D., 1996. Finite element analysis of quasi-static and dynamic collapse behaviors of framed structures by the adaptively shifted integration technique. Computers and Structures 58 (5), 947–955.

Toi, Y., Isobe, D., 2000. Analysis of structurally discontinuous reinforced concrete building frames using the ASI technique. Computers and Structures 76 (4), 471–481.

Website of ASIFEM3: Finite element code for framed structures using ASI technique (Cubic beam element version). http://www.kz.tsukuba.ac.jp/~isobe/asifem3_e.html.

APPENDIX C

Ceiling Collapse Analysis of a Gymnasium

C.1 INTRODUCTION

In this appendix, another application example of the ASI-Gauss code, besides collapse analysis of buildings, is discussed. Fig. C.1 shows an example of ceiling collapse damage occurred inside a gymnasium during the 2011 Great East-Japan Earthquake. One can observe the plaster boards, weighing about 10 kg each, falling and scattering all over the floor. The prevention of these phenomena is an important issue not only to save people's lives, but to keep these facilities as safety shelters after earthquakes. A numerical analysis to simulate the ceiling collapse in a full-scale gymnasium specimen, tested at the E-Defense shaking-table facility in 2014, was conducted (Isobe, 2017).

Figure C.1 Ceiling collapse damage observed in a gymnasium after the 2011 Great East-Japan Earthquake (NILIM & BRI, 2012).

C.2 COMPONENTS OF SUSPENDED CEILINGS AND THEIR STRENGTHS

The components of the suspended ceilings are quite complicated as shown in Fig. C.2. The clips, indicated in red circles in the figure, connect ceiling joists to their receivers. These are small and delicate components and can be detached during repeated excitation. Once there is a local detachment of clips, a change in the load distribution may cause a chain reaction of detachments, which ends in a drop of plaster boards. Furthermore, the detachments of hanging bolts that are connected to the structural members composing the roof, and failure of screws on plaster boards are assumed to be other main causes of the ceiling collapse.

There were some preliminary tests conducted to see the actual strengths of these components as shown in Fig. C.3. For example, the opening of clips began at a tensile force of 0.4 kN (Nakagawa and Motoyui, 2006) and the opening of hangers at 2.8 kN (Sugiyama et al., 2009). For the screws connecting plaster boards to ceiling joists, they were observed to be shear damaged at a value of 0.3 kN (Sugiyama et al., 2010), and their heads became loose when 0.2 kN of tensile force was applied (Sakurai et al., 2009). These criteria were implemented in the analysis; however, the tensile strengths of clips were overestimated and larger criteria of 0.8 kN was adopted here, because they tend to slip on ceiling joists during excitation.

Figure C.2 Configuration of a suspended ceilings.

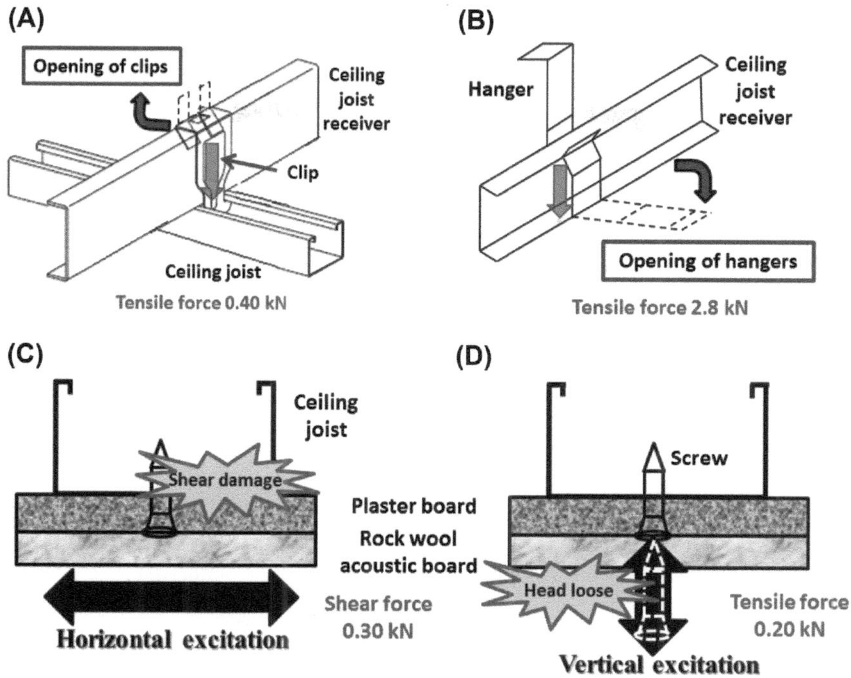

Figure C.3 Main causes of the ceiling collapse and their detachment criteria. (A) Opening of clips, (B) opening of hangers, (C) shear damage of screws, and (D) head loose of screws.

C.3 NUMERICAL MODELING OF CEILINGS

Fig. C.4 shows a ceiling model in detail. As the ASI-Gauss code was applied, all components (except braces and hanging bolts) were modeled with two linear Timoshenko beam elements per member. Hangers and hanging bolts were modeled in one piece. The plaster boards were assumed as rigid in outer plane direction and only the mass of rock wool boards was considered. Their strengths were neglected. Clips and screws were modeled with minute, small elements. Each plaster board was modeled separately to consider local contact between plaster boards; it was realized by modeling the screws slightly apart. Fig. C.5 is a constructed gymnasium model with the ceilings. Elasto-plastic buckling of braces and hanging bolts were taken into account by modeling them with eight beam elements each and two hinge elements on both ends.

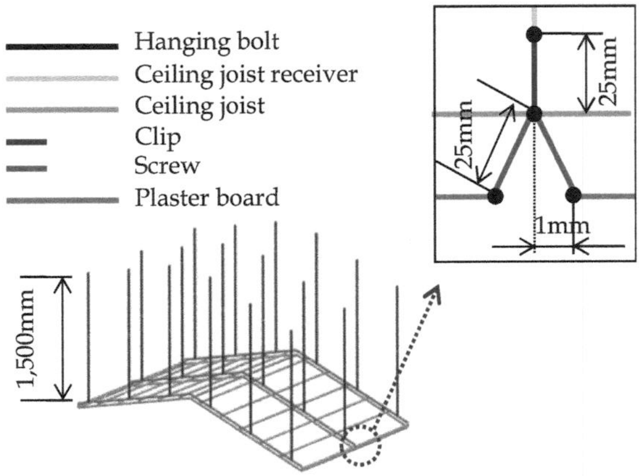

Figure C.4 Details of a ceiling model.

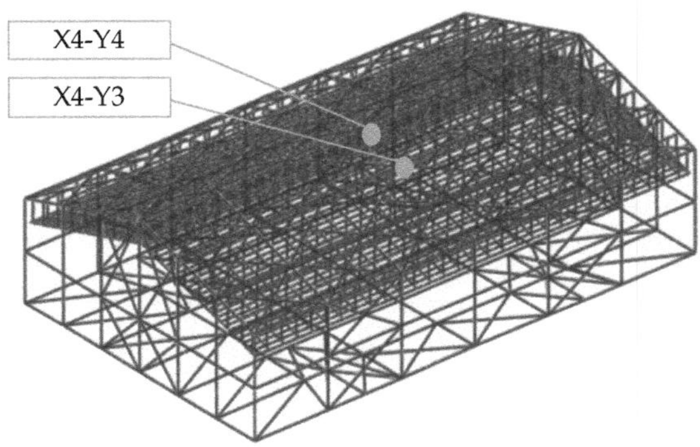

Figure C.5 Gymnasium model with the ceilings.

C.4 CEILING COLLAPSE ANALYSIS OF A GYMNASIUM

The numerical result was validated with the experimental result performed at the E-Defense under a 50% input wave of K-NET Sendai. The acceleration data obtained on the shake-table floor of the E-Defense, as shown in Fig. C.6, was used as an input wave for the analysis.

The acceleration responses and spectrum obtained at the location (X4-Y3) on the roof indicated in Fig. C.5, for example, match well with the experimental result as shown in Figs. C.7 and C.8. A 10-Hz low-pass filter

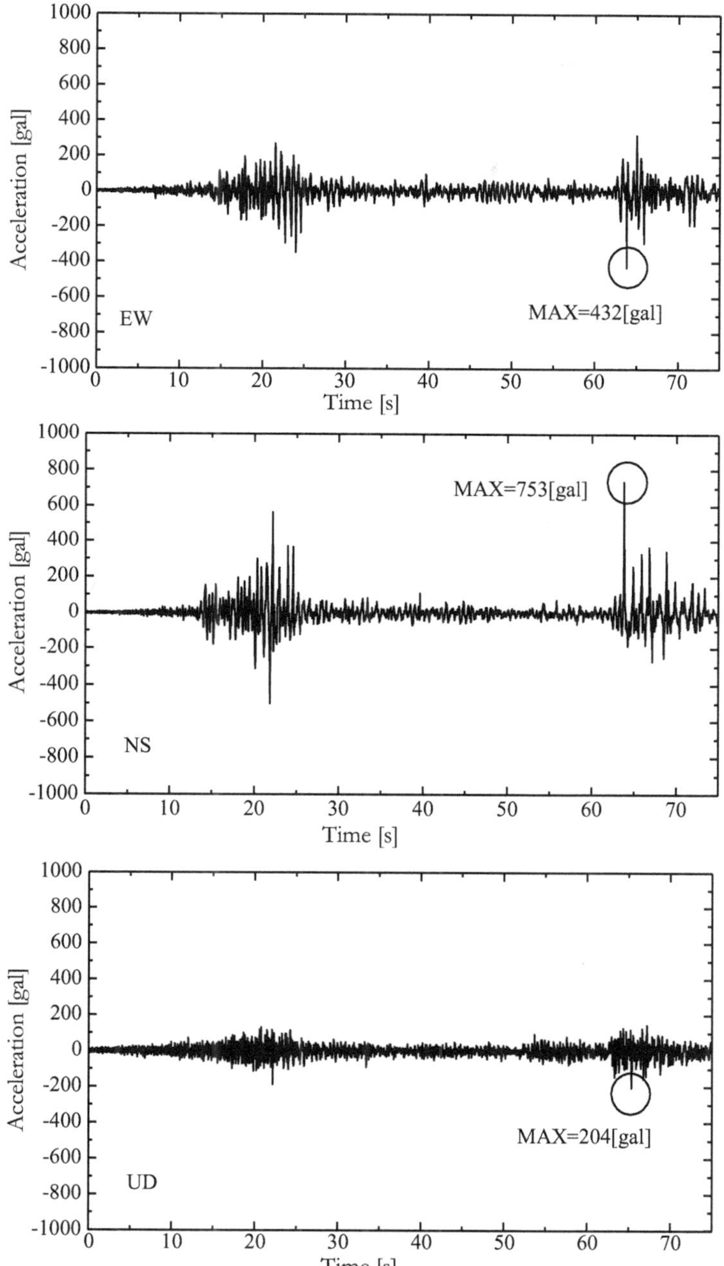

Figure C.6 Input ground acceleration (obtained data on shake table when K-NET Sendai 50% wave was applied).

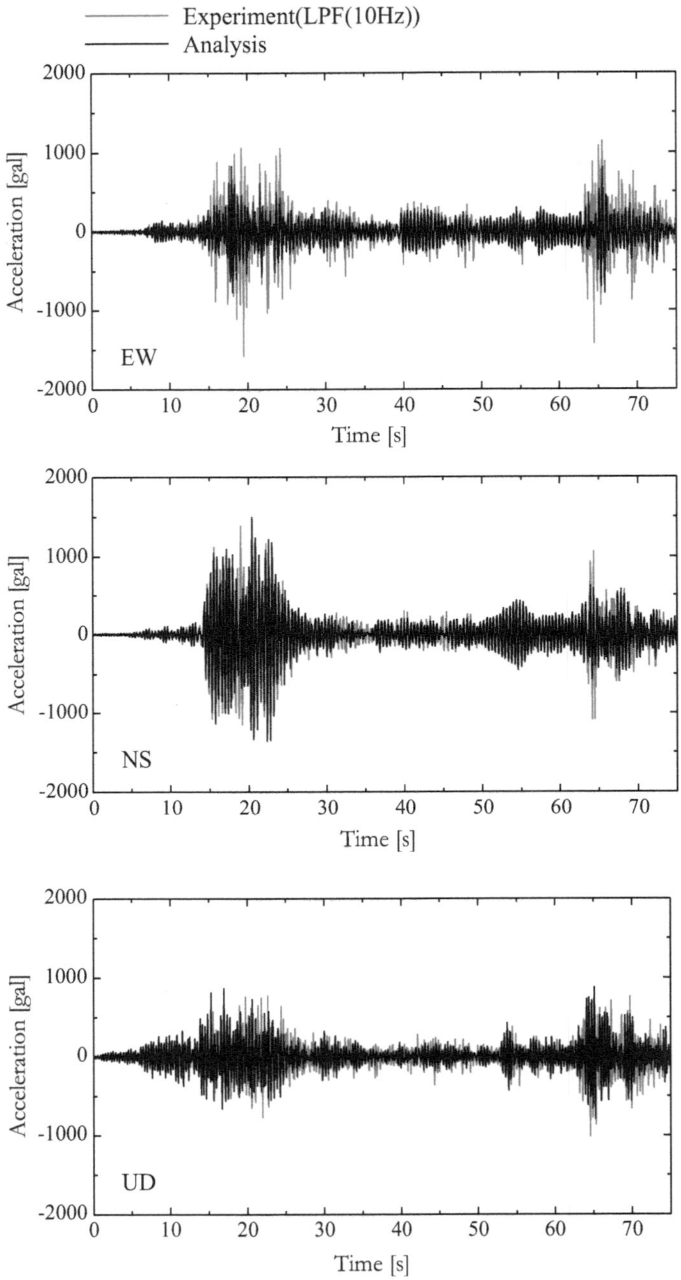

Figure C.7 Acceleration responses (X4-Y3).

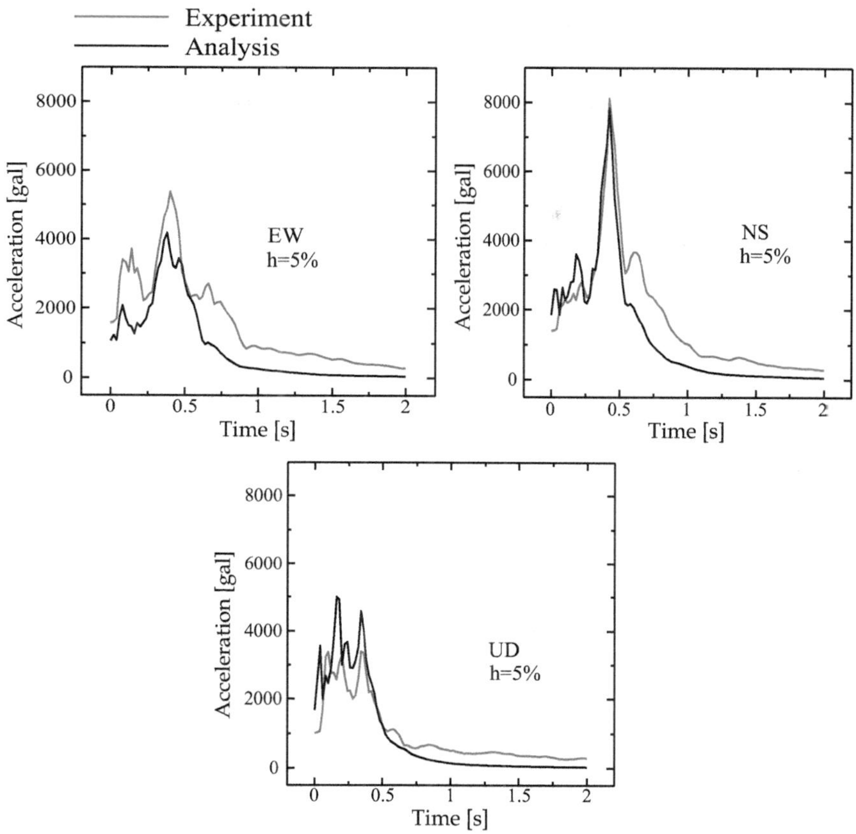

Figure C.8 Acceleration-response spectrum (X4-Y3).

was used in the post-processing procedure of the data. The displacement responses obtained at the location (X4-Y4), for example, also match well with the experimental results as shown in Fig. C.9. The phase and the amplitude agrees well as shown in Fig. C.10 in the extracted responses during 15—25 s, the first peak of the input wave.

Figs. C.11 and C.12 show the collapse behavior of the ceilings from the front view and the global view of the gymnasium. The plaster boards near walls pattered down occasionally at the first peak of the input wave. These are due to detachment of clips and screws caused by collisions to the walls.

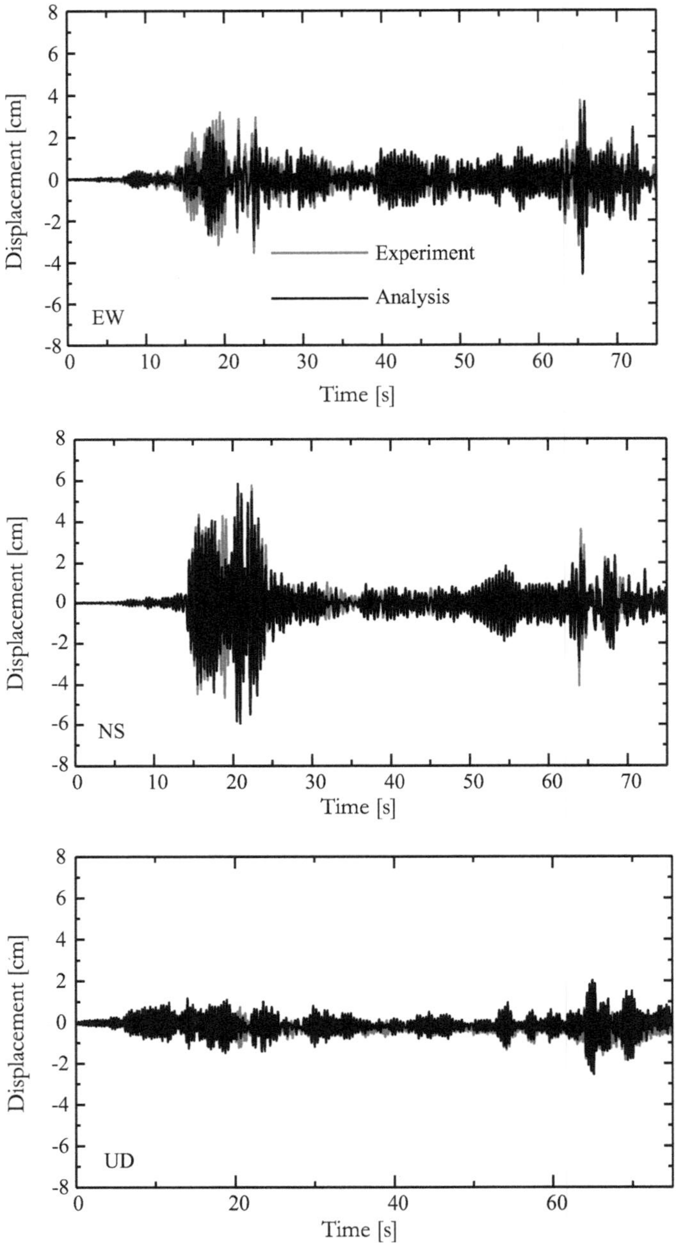

Figure C.9 Displacement responses (X4-Y4).

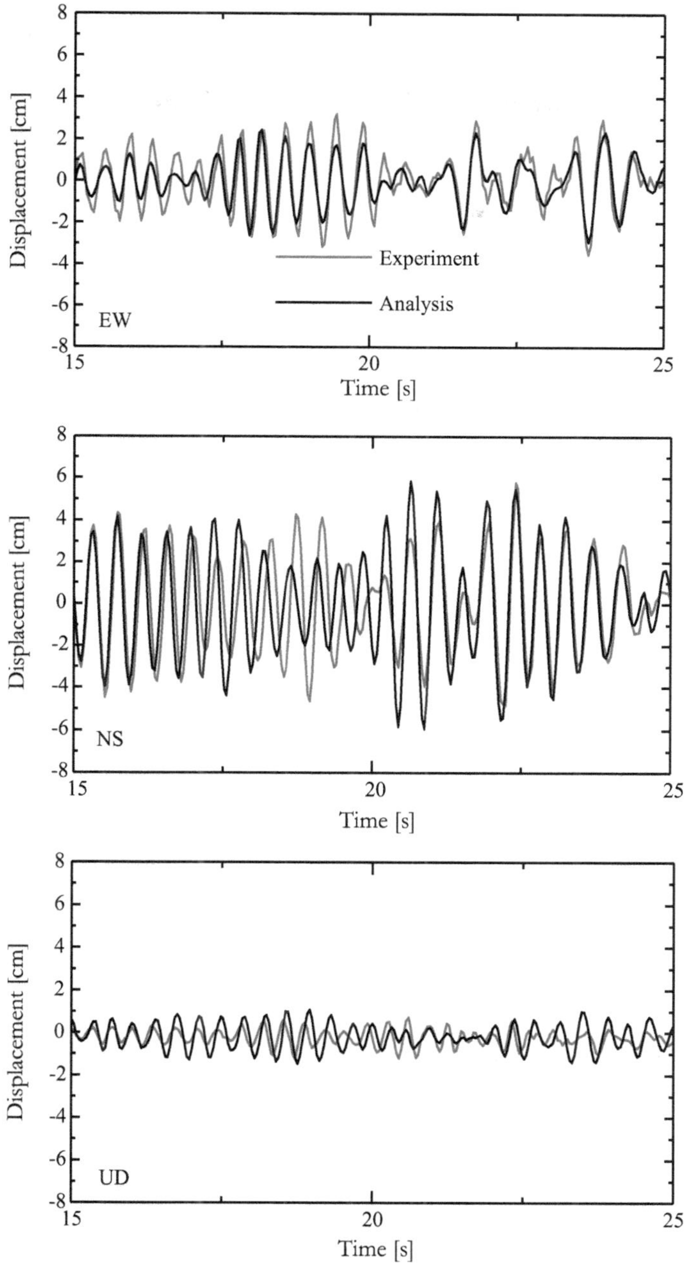

Figure C.10 Displacement responses during 15–25 s (X4-Y4).

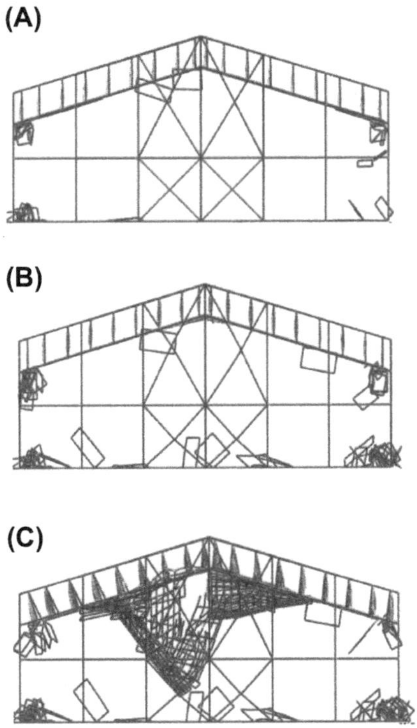

Figure C.11 Collapse behavior of the ceilings (front view). (A) 17.6 s, (B) 59.0 s, and (C) 72.9 s.

The clips near roof top began to get loose due to buckling of hanging bolts caused by vertical excitation, which ends, at the second peak of the input wave, in drop of plaster boards in a wide range.

C.5 SUMMARY

As shown in this appendix, the ASI-Gauss code can also be applied to the motion behavior analysis of nonstructural components such as the ceilings. It is expected to gain specific information of the collapse mechanism and the means to prevent the collapse of ceilings through a series of simulations. Another application of the code on the motion behavior analysis of furniture is shown in Appendix D.

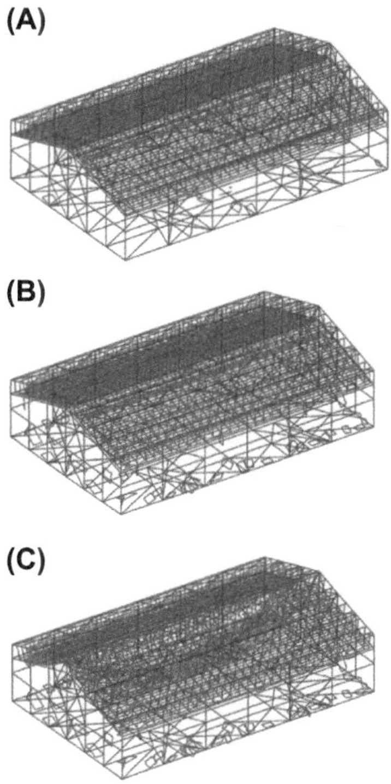

Figure C.12 Collapse behavior of the ceilings (global view). (A) 17.6 s, (B) 59.0 s, and (C) 72.9 s.

REFERENCES

Isobe, D., 2017. Simulating collapse behaviors of buildings and motion behaviors of indoor components during earthquakes. In: Zouaghi, T. (Ed.), Earthquakes – Tectonics, Hazard and Risk Mitigation. InTech, ISBN 978-953-51-2885-4, pp. 341–363.

Nakagawa, Y., Motoyui, S., 2006. On mechanical characteristics of connection parts in ceilings. In: Summaries of Technical Papers of Annual Meeting AIJ 2006, B-1, pp. 845–846 (in Japanese).

NILIM & BRI, 2012. Report on Field Surveys and Subsequent Investigations of Building Damage Following the 2011 Off the Pacific Coast of Tohoku Earthquake. National Institute for Land and Infrastructure Management & Building Research Institute.

Sakurai, S., Kumagai, S., Nagai, T., Kawaguchi, K., Ando, K., Araya, M., 2009. Fundamental study of non-seismic failure of suspended ceilings of swimming pools. In: Summaries of Technical Papers of Annual Meeting AIJ 2009, B-1, pp. 897–898 (in Japanese).

Sugiyama, T., Kashiwazaki, T., Kobayashi, T., Nukui, Y., Yabuuchi, A., 2009. Static loading tests on structural elements and full-scale partial model of conventional type ceiling. In: Summaries of Technical Papers of Annual Meeting AIJ 2009, B-1, pp. 227–228 (in Japanese).

Sugiyama, T., Kashiwazaki, T., Nosohara, M., Nukui, Y., Suzuki, A., 2010. Shear tests of screwed joint between ceiling joist and plaster board of conventional type ceiling. In: Summaries of Technical Papers of Annual Meeting AIJ 2010, B-1, pp. 871–872 (in Japanese).

APPENDIX D

Motion-Behavior Analysis of Furniture During Earthquakes

D.1 INTRODUCTION

In high-magnitude earthquakes, improperly secured furniture could become dangerous for human life even if the building itself had no damages at all. It is crucial especially on the upper floors of high-rise buildings excited under long-period ground motion. Many tumbled furniture, such as chairs and tables in schools, could also become fatal obstacles that obstruct children from evacuating. Thus, it is important to examine the motion behaviors of furniture, as well as the collapse behaviors of the building, to save people's lives during big earthquakes. The E-Defense of NIED carried out several experiments to investigate the motion behaviors of furniture under various conditions using a three-dimensional full-scale shaking table and obtained useful information (E-defense, 2009). In parallel, the E-simulator project is aiming to develop an effective simulator for investigating the motion behaviors of such furniture under a broader set of conditions. The discontinuous element method (DEM) (Cundall, 1971) can be useful for this purpose; however, the mechanical behaviors of the constituent materials of the structural and nonstructural components may not be well examined. In this appendix, the results of motion-behavior analyses of furniture conducted using the ASI-Gauss code with a contact algorithm based on sophisticated penalty method are shown. The results are then compared with the experimental results performed on a shake table (Isobe et al., 2015; Isobe, 2017).

D.2 CONTACT ALGORITHM USING SOPHISTICATED PENALTY METHOD

To simulate various contact phenomena between furniture and walls or floors during seismic excitation, frictional contact between objects was considered using a sophisticated penalty method (Isobe et al., 2012, 2015;

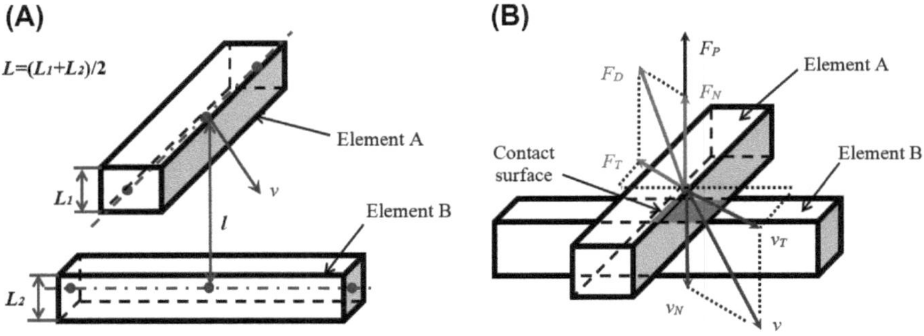

Figure D.1 Subjected forces and the geometrical relations between two elements. (A) Geometrical relations between the approaching elements and (B) forces subjected at the contact plane.

Ogino et al., 2015). Fig. D.1 shows the subjected forces and the geometrical relations between two elements when they approach each other with a relative velocity $\mathbf{v}$. Once the current distance l between the central lines of two elements becomes shorter than the mean value L of the member widths of the elements, the penalty force vector $\mathbf{F_P}$, expressed as follows, is assumed to act in the normal direction of the contact plane.

$$\mathbf{F_P} = \alpha\left(1 - \frac{l}{L}\right)^q \frac{\mathbf{n}}{\|\mathbf{n}\|} \quad \text{if } l \leq L \tag{D.1}$$

where α is the penalty coefficient, q the penalty index, and $\mathbf{n}$ the normal vector at contact surface. Then, the frictional force vector $\mathbf{F_D}$ is assumed to act not only in the tangential direction of the contact plane but also in the normal direction; the frictional force vector is decomposed into the following two components.

$$\mathbf{F_T} = -\mu\alpha\left(1 - \frac{l}{L}\right)^q \frac{\mathbf{v_T}}{\|\mathbf{v_T}\|} \quad \text{if } l \leq L \tag{D.2a}$$

$$\mathbf{F_N} = -D_C\left(1 - \frac{l}{L}\right)^q \frac{\mathbf{v_N}}{\|\mathbf{v_N}\|} \quad \text{if } l \leq L \tag{D.2b}$$

where μ is the dynamic friction coefficient, D_c the coefficient related to damping for normal direction, $\mathbf{v_T}$ the tangent component of $\mathbf{v}$, and $\mathbf{v_N}$ the normal component of $\mathbf{v}$. The component in the normal direction $\mathbf{F_N}$ acts as a damping force along the contact depth and contributes to maintain numerical stability.

One of the features of the developed code is that it is easier to select the contact parameters as compared to the DEM. The penalty coefficient and dynamic friction coefficients were fixed based on quite simple rules. First, the penalty coefficient α was fixed as the weight of each furniture. Then, the dynamic friction coefficients were fixed as 80% values of each static friction coefficient, which were measured before the experiments on the shake table. As for the coefficient related to damping D_c, they were fixed as 120% values of the penalty coefficient α. These rules were applied throughout all the analysis.

D.3 MOTION-BEHAVIOR ANALYSIS OF FURNITURE

A set of furniture, shown in Fig. D.2, was used for the numerical analyses and the experiments. It included total five items from a heavy cabinet of about 44 kg to a light office chair with casters of about 6 kg. The sizes and locations of center of gravity of each item are shown in Table D.1.

Fig. D.3 shows the arrangement of furniture on a shake table and their finite element models constructed using linear Timoshenko beam elements. The motions of each item were measured by using a motion capture system with six infra-red cameras arranged around the shake table. The contact parameters for each item were set as shown in Table D.2. These parameters were determined according to the rules described earlier.

Fig. D.4 shows the state of furniture at the end of different input waves. The motions of each item, particularly the rocking motions of the cabinets and the sliding motions of those with casters, are well simulated. Fig. D.5 shows the comparison of displacements in X-direction at the top of the separated cabinets, as an example. For all the cases of different input waves, both results match almost perfectly.

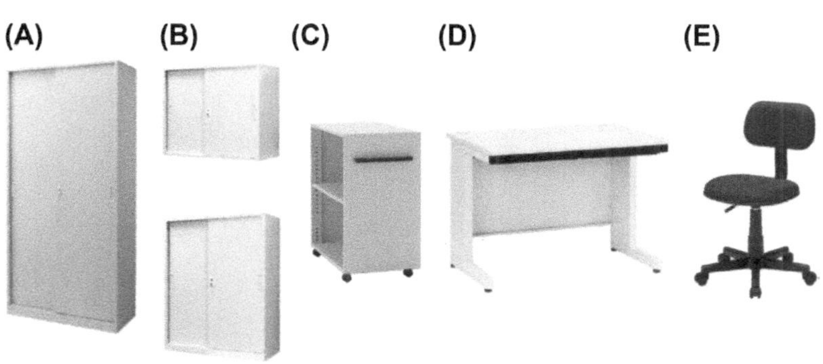

Figure D.2 A set of furniture used for numerical analyses and experiments. (A) Tall cabinet, (B) separated cabinets, (C) side table, (D) desk, and (E) office chair.

Table D.1 Sizes and Locations of Center of Gravity of Each Furniture

Furniture	Size (mm)			Center of Gravity (mm)			Mass (kg)
	H	*D*	*W*	*h*	d_1	d_2	
A	1850	400	900	925	178	222	43.6
B-upper	730	400	900	365.5	171	229	21.3
B-lower	1120	400	900	562	179	221	28.8
C	617	340	550	308	218	122	8.0
D	700	700	1200	600	380	320	30.5
E	810	520	520	423	260	260	6.2

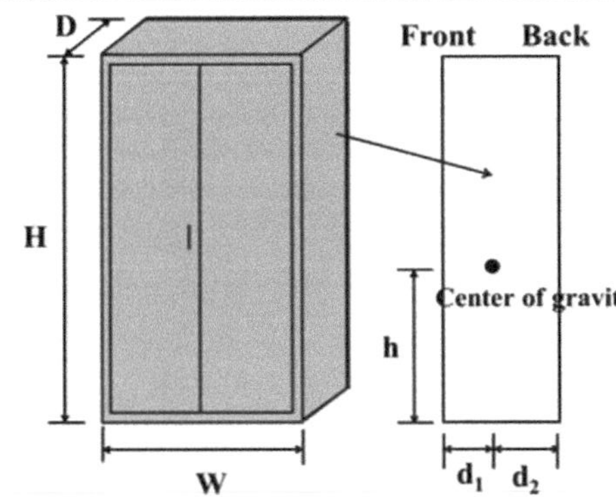

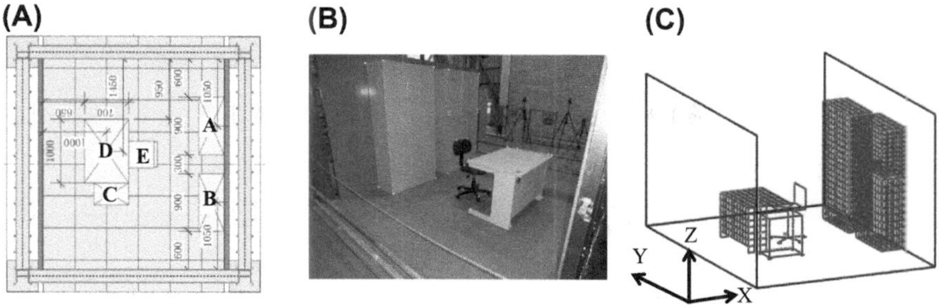

Figure D.3 Arrangement of furniture on a shake table and their finite-element models. (A) Arrangement of furniture, (B) furniture on shake table, and (C) finite-element models.

Table D.2 Contact Parameters for Each Furniture

Furniture	Penalty Coefficient α (kgf)	Coefficient Related to Damping D_c (kgf)	Dynamic Friction Coefficient μ			
			vs. Floor		vs. Wall	vs. Furniture
			Long Side	Short Side	–	–
A	43.6	52.3	0.278	0.208	0.278	0.169
B-upper	21.3	25.6	0.169	0.169	0.169	
B-lower	28.8	34.6	0.230	0.270	0.230	
C	8.0	9.6	0.144	0.094	0.144	
D	30.5	36.6	0.377	0.354	0.377	
E	6.2	7.4	0.065	0.038	0.065	

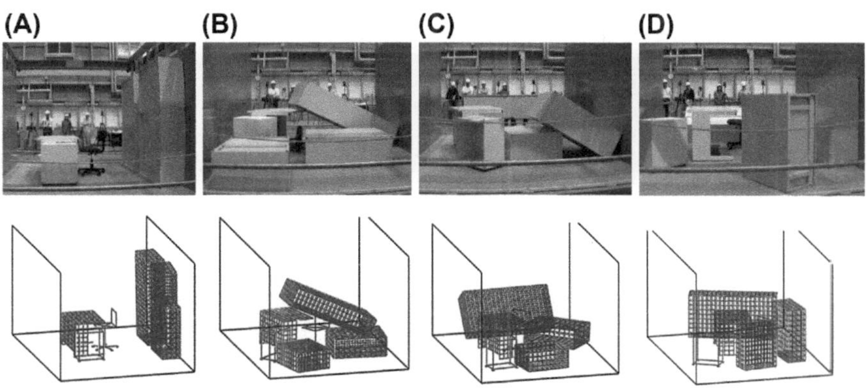

Figure D.4 State of furniture at the end of different input waves (upper: experiment, lower: analysis). (A) JMA-Kobe 1-D, (B) JMA-Kobe 1-D, (C) JMA-Kobe 3-D, and (D) KiK-net Haga.

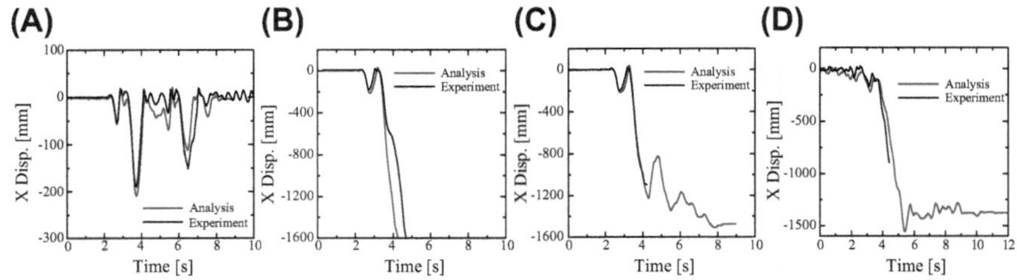

Figure D.5 Comparison of displacements in *X*-direction at the top of the separated cabinets. (A) JMA-Kobe 1-D, (B) JMA-Kobe 1-D, (C) JMA-Kobe 3-D, and (D) KiK-net Haga.

D.4 SUMMARY

The numerical code and the frictional contact algorithm shown in this appendix succeeded in simulating the motion behaviors of furniture subjected to various seismic excitations. One of the advantages of applying the finite-element approach is that, though it is not clearly shown in this appendix, it enables one to evaluate the deformations and sectional force distributions in furniture as well as their motion behaviors.

REFERENCES

Cundall, P.A., 1971. A computer model for simulating progressive, large-scale movement in blocky rock system. In: Proc. Int. Symp. on Rock Mechanics, Nancy, France, 1971; II-8, pp. 129–136.

E-defense, 2009. National Research Institute for Earth Science and Disaster Prevention. http://www.bosai.go.jp/hyogo/index.html.

Isobe, D., 2017. Simulating collapse behaviors of buildings and motion behaviors of indoor components during earthquakes. In: Zouaghi, T. (Ed.), Earthquakes – Tectonics, Hazard and Risk Mitigation. InTech, ISBN 978-953-51-2885-4, pp. 341–363.

Isobe, D., He, Z.H., Kaneko, M., Hori, M., 2012. Motion analysis of furniture under sine wave excitations. In: Proceedings of the International Symposium on Reliability Engineering and Risk Management (ISRERM2012), pp. 230–233. Yokohama, Japan.

Isobe, D., Yamashita, T., Tagawa, H., Kaneko, M., Takahashi, T., Motoyui, S., 2015. Motion analysis of furniture under seismic excitation using FEM. Journal of Structural and Construction Engineering AIJ 80 (718), 1891–1900 (in Japanese).

Ogino, H., Yamashita, T., Kaneko, M., Isobe, D., 2015. Development of a finite element code to simulate behaviors of furniture under seismic excitation. Journal of Structural and Construction Engineering AIJ 80 (717), 1687–1697 (in Japanese).

INDEX

'*Note*: Page numbers followed by "f" indicate figures and "t" indicate tables.'

N

P

R

S

FSC
www.fsc.org
MIX
Papier | Fördert gute Waldnutzung
FSC® C083411

Kolibri 360 GmbH
Ferdinand-Jühlke-Straße 7
99095 Erfurt, Deutschland
produktsicherheit@kolibri360.de